ENERGIZING INDIA

ENERGIZING INDIA

Towards a Resilient and Equitable Energy System

Suman Bery
Arunabha Ghosh
Ritu Mathur
Subrata Basu
Karthik Ganesan
Rhodri Owen-Jones

Los Angeles | London | New Delhi
Singapore | Washington DC | Melbourne

First published in 2016 by

SAGE Publications India Pvt Ltd
B1/I-1 Mohan Cooperative Industrial Area
Mathura Road, New Delhi 110 044, India
www.sagepub.in

SAGE Publications Inc
2455 Teller Road
Thousand Oaks, California 91320, USA

SAGE Publications Ltd
1 Oliver's Yard, 55 City Road
London EC1Y 1SP, United Kingdom

SAGE Publications Asia-Pacific Pte Ltd
3 Church Street
#10-04 Samsung Hub
Singapore 049483

Published by Vivek Mehra for SAGE Publications India Pvt Ltd.

Library of Congress Cataloging-in-Publication Data Available

ISBN: 978-93-859-8523-2 (HB)

Contents

Figures

Tables

Boxes

Foreword

A global energy transition is underway as the world seeks to meet rising demand while lowering carbon emissions. Yet energy systems are large, complex and slow to change. Decisions taken today will continue to have an impact well into the middle of the century.

The energy choices that India makes will be critical to meeting its development goals. They will also be influential outside its borders. That is why researchers from three institutions joined forces three years ago: to explore the energy choices available to India's society and economy at a time of considerable uncertainty, but also opportunity, in the global energy system.

This publication is a distillation of our joint work. It draws on the extensive knowledge of the three partners: Shell's scenarios group based in The Hague, The Netherlands; the Council on Energy, Environment and Water (CEEW); and The Energy and Resources Institute (TERI), both based in New Delhi. It also draws on analysis and modelling undertaken as part of this collaboration.

This book is not a report on the research findings. Nor is it designed solely for use by the government. Instead, the motivation to write and publish the book stemmed from our belief that national energy choices around the world, while undoubtedly technically challenging, are ultimately also political.

As such, informed and sustainable choices require active public discussion, particularly in India, a society with a well-established tradition of lively public debate. As a key driver of India's economic growth, and as a significant determinant of broader human progress, India's energy choices deserve widespread discussion—within universities, government, industry, research institutions and the media. The purpose of this book is to stimulate such a conversation.

Finally, the scope of our partnership is important and worthy of recognition. Solving the challenge of providing more and cleaner energy will require institutions around the world to operate outside their comfort zones. This principle applies as much to research institutions as to government departments.

The intellectual partnership, between a global energy multinational and two respected Indian research institutions, was a learning experience for all three partners and demonstrates the power of cross-border and cross-disciplinary collaboration. We congratulate this effort.

Harry Brekelmans
Projects & Technology Director, Royal Dutch Shell

Jamshyd N. Godrej
Chairperson, Council on Energy, Environment and Water

Ajay Mathur
Director-General, The Energy and Resources Institute

Preface

This book explores the opportunities and challenges in articulating and implementing a resilient, robust but flexible set of strategies for meeting India's primary energy needs. It has been written at a time of important transitions and considerable uncertainty for both India and the world. India is now among the world's fastest growing large emerging market economies. It is also by far the poorest member of the G-20, as measured by real per capita GDP, and both less industrialized and less urbanized than its peers.

The book deals with the energy system seen as a whole rather than following a sector approach. While several recent government reports have also followed this approach, we have two specific reasons for doing so: to develop a sense of what matters most in helping move the energy system towards meeting agreed long-term goals, and to identify common issues that span individual fuels or users. Accordingly, the book has been organized by cross-cutting themes rather than by energy source or user sector. While attention is paid to both the demand and supply sides of the energy system, the dimension given the most attention is the primary energy mix, which most directly affects both energy security and environmental sustainability.

Intelligent energy policy requires an informed public debate on the energy choices and uncertainties facing India. In all countries, energy choices reflect economic, domestic, political and geopolitical considerations. India is no exception, and there can be no single correct answer. Cost and distributional implications are looked at to provide a thorough analysis of the trade-offs that different policy options present, and to identify the costs and trade-offs associated with one path versus another. Wherever possible the effort is to identify sweet spots that minimize avoidable costs.

The book does not attempt to be directive or prescriptive with false precision. It reflects the work of three institutions, two based in India (CEEW and TERI) and one abroad (Shell). The partnership combines informed views on global long-term trends in energy and energy technology with detailed local knowledge of Indian issues. One of the three collaborating institutions (Shell) has an established tradition of addressing uncertainty through the use of scenarios. Accordingly, a range of scenarios is explored both qualitatively and, where possible, quantitatively. There is value in taking a medium- and long-term view as the lifetime and gestation of much infrastructure is in decades as opposed to months. Since many of the uncertainties have to do with global developments in supply, regulation and technology, this report makes use of Shell's global energy scenarios published in 2013, the New Lens Scenarios. The hope is that this book can make a modest contribution to a more informed national debate on India's energy options. For this reason as well, while the participating institutions have given encouragement and support to the researchers involved, the views presented in this volume are those of the authors even as they draw upon past work of their parent institutions.

Acknowledgements

The authors would like to thank the following for their varied contributions in making this book possible:

Harry Brekelmans, Matthias Bichsel and Jeremy Bentham at Shell for driving this initiative. Renana Jhabvala, Nitin Desai, N.K. Singh, Kirit Parikh, Shyam Saran, T.N. Ninan, Yasmine Hilton and Leena Srivastava for their guidance to the authors at numerous points.

Saptarshi Das, Ilika Mohan, Aayushi Awasthy, Anjali Ramakrishnan, Swati Mitchelle D'Souza, Madhura Joshi, Siddharth Singh for actively working and Prabir Sengupta at TERI, Russel Faulkner at Shell, Shalu Agrawal and Vaibhav Gupta at CEEW for contributions to individual chapters.

Nort Thijssen, Jerome de Morant, Shoba Veeraraghavan and Sharvani Thakur at Shell for design and support of the India Energy Model (INEM) and its transfer to India.

Bruce Ross-Larson at Communications Development Inc. for editorial guidance.

Amit Soman, Neeti Tandon, Nitin Prasad and Rajesh Ganesh at Shell for reviewing and commenting on the draft at various stages.

The Rockefeller Foundation's Bellagio Center where some of the ideas on energy security were researched and developed by CEEW.

Energizing India

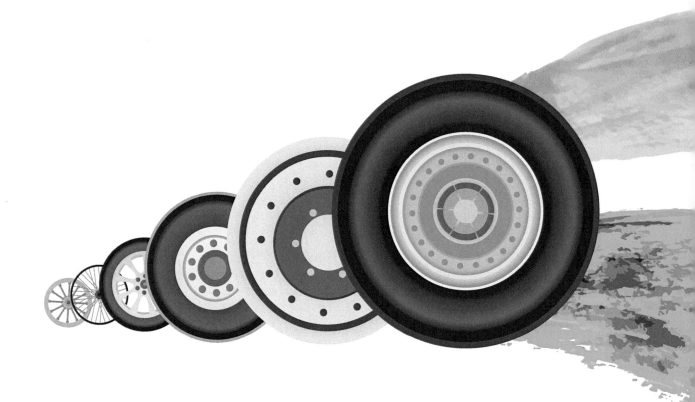

With a population of about 1.25 billion and a purchasing power parity (PPP) GDP per capita of roughly $5,700, India is in the midst of a huge transformation as the economy grows, incomes rise, manufacturing becomes a bigger part of the economy and the country becomes more urbanized. This growth will drive a sustained expansion of infrastructure for energy, urban development and transportation.

India's modern energy system is already sizable, with the world's third largest electricity generation capacity. Rapid economic growth and a rising share of manufacturing in GDP will underpin growing demand for modern energy. About 75 million households, a third of the total, are not connected to grid electricity, and 75–80% of rural households use traditional biomass as a primary source for cooking. Improving their energy access will further enhance India's future demand for modern energy. And given the very low energy consumption per capita, the aspirations for a more energy-intensive lifestyle will almost certainly grow.

India's energy choices are likely to be conditioned by less ample carbon space than was available to countries whose development preceded it—and by its vulnerability to climate-related stress. India could however benefit from the considerable research and innovation taking place globally on cleaner energy sources, based on which it could leapfrog to cleaner and more efficient options even faster.

India's energy system has been in a process of steady evolution for the last two decades, particularly in blending private involvement in what previously had been a largely state-dominated sector. This process will continue and indeed can be expected to accelerate, raising fresh challenges of coordination and communication. Coordination between India's states and the Union Government is a well-established feature of India's vibrant federal polity, and this cooperation will face new challenges with the current government's commitment to competitive federalism.

Four impending transitions

The formidable challenges ahead are best articulated as a series of transitions from the situation today to a desired end-state, in the presence of continuous change. The biggest and, in many ways, most difficult shift will be in moving from traditional to modern sources of energy. Rising incomes have not yet translated into higher consumption of modern energy in many parts of India, despite long-standing efforts to provide income support, extend the grid, use the public distribution system to provide subsidized energy and introduce modern cook-stoves and solar lanterns. In a democracy with a large young population, the drive to move away from traditional fuels will be inexorable.

Another big shift will be from rural energy demand to urban. India is still predominantly rural, and much of the deficit in energy access is particularly evident in rural areas. In the near term, meeting basic energy demand in rural India is a pressing priority. But the growth of the urban population will be much faster than that of rural, bringing not only new challenges but also opportunities. By 2050, 38% of Indians will be city dwellers. This will have implications not just for basic energy access—it will also drive energy demand in new sectors: buildings, air conditioning, transportation (public and private) and commerce. Meanwhile, energy demand in rural areas will also rise, as more energy is needed for a range of productive activities, such as agricultural mechanization, food processing and small-scale rural industries.

A third shift will be deeper integration into global energy markets. India has relied on imported crude oil for a long time. But the largest sources of energy—coal and traditional biomass—have been for the most part domestically procured. This allowed India to be fairly autarkic in its approach to energy policy. But in the 21st century India is on track to assuming a role in global energy trade similar to Europe's in the 20th century. It will not be the biggest energy consumer, but it can no longer play only at the margins. India will be a "swing voter" in global energy markets, with a strong national interest in well-functioning markets.

Meanwhile, the global energy system is going through its own perturbations. This is most evident in the sharp decline in the dollar price of oil since 2014, reflecting robust production from unconventional sources in North America, and in the commercial response from members of OPEC. No definitive conclusion can as yet be drawn on the longer term implications, except that price volatility will continue to be a feature of global oil markets, and such volatility needs to be factored into Indian planning and decision-making. Natural gas as an energy resource is increasingly global and is no longer limited by cross-border pipeline infrastructure. The pattern of global coal production, trade and use is being reshaped by declining demand in the United States and in China.

These shifts—driven by the changing structure of the Indian economy—are happening against the backdrop of climate change. Depending on the level of global emissions, projections of global temperature rising above pre-industrial levels between 2°C and 4°C by the century's end could lead to severe impacts on water, food, coastal flooding, heat stress and health impacts—as well as systemic impacts on international security. The choices India makes and the pace of shifting to a lower carbon economic pathway have a bearing on the fourth transition of environmental sustainability.

The sharp fall in the cost of renewable technologies—such as solar photovoltaics (solar PV) as a result of market and policy-driven decarbonization efforts—provides alternatives to the fossil fuel sources that have been the engine of economic progress over the last two centuries. This transition is already under way in many countries, but the potential for scaling up its implementation in countries like India is again contingent on technological progress, prices and policies. Must the more intensive industrialization phase awaiting India be fuelled largely by oil, coal and gas? Or do new technologies provide opportunities for leapfrogging dependence on fossil fuels at an acceptable cost?

India's energy aspirations

India's energy choices will be driven by several key goals. Reliable modern energy is essential in a modern society and indispensable for enhancing economic opportunity and the quality of life. For this reason, one of the most important goals should be to infuse technology and to design competitive market structures that will quickly ensure universal physical and economic access to economically viable modern energy alternatives for all residents, particularly those with limited means and those living in remote areas. The success of Indian mobile telephony over the last 15 years provides an encouraging precedent.

Equally, the scope of ambition for energy access should go beyond lifeline levels of residential energy (say in the form of a couple of LED bulbs for lighting or a cleaner stove for cooking). The impact of constrained and unreliable access to energy is manifested through many channels: health and educational outcomes, the time value of money in collecting traditional energy sources, or an overall decline in household-level productivity, entrenching many in an energy and income poverty trap. Thus, a baseline level of access to productive energy should be also recognized as part of India's energy aspirations. Productive energy, say for small-scale agricultural use or to empower communities to adopt additional income generating activities, is essential if the link between energy access and human development progress has to be maintained. In short, energy first and foremost is an issue of both lives and livelihoods.

A second goal is to ensure reliability and resilience of the entire energy system so that economic activity is shielded from disruptions arising from the energy sector. The aspect that features most prominently in policy discussions is India's vulnerability to imported energy, but disruptions can arise from domestic sources as well, and addressing them is of equal importance. This resilience will be driven by advances in technologies and related costs, opportunities to change the fuel mix, and the policies to expand energy infrastructure on the supply side as well as to shape behaviour and consumption patterns on the demand side. The reliability of the energy system will be judged against its ability to contribute to rapid economic growth for India in the coming decades.

A third goal is the long-term sustainability of the energy system in various dimensions. This most obviously includes both environmental sustainability and personal health and safety at all stages in the energy production and consumption chain. One element in environmental sustainability is India's recognition of a global carbon constraint, but this is far from being the only one. Local air pollution, indoor air pollution, environmental and safety concerns surrounding fuel production including land and water management issues—all fall under this head, what might loosely be called sustaining the energy industry's "license to operate."

Sustainability also has an important fiscal and financial dimension. Resources have to be priced to recover costs and encourage investments. And subsidies need to be clearly motivated and targeted, with objective review milestones and exit strategies in mind.

How will India manage the four transitions—from traditional to modern energy, from rural to urban areas, from autarky to global integration and from a high-carbon to a low-carbon but robust economy? Will it be ad-hoc, incremental and arbitrary? Or will India's strategy and policies demonstrate a coordinated approach across government departments and between the government and the private sector? Will India plan for the long term while sequencing a series of short-term milestones in developing its energy choices? And will this approach be strategic, implying recognition that energy policy will be a core determinant of its economic transformation over the course of this first half-century? Managed well, India can demonstrate that a rapid shift to modern energy can be compatible with social, economic and environmental sustainability. Managed poorly, the true potential of India's role and influence in the global economy will remain unrealized. The stakes are, therefore, substantial.

If this book had been written in, say, the early 1970s, the main criterion for making energy choices would possibly have been only cost. However, at this juncture, planning for energy choices is increasingly tilted towards properly reflecting the opportunity costs of finite resources, and health costs attributable to pollution from conventional fossil fuels. Moreover, today, there is global recognition of the fact that the carbon constraint is real. So cost, time scales and sustainability across multiple dimensions will all need to be relevant filters to evaluate India's energy choices.

As the four transitions are under way, India's energy policies would have to satisfy some basic imperatives from a political perspective: greatly accelerating the pace of reducing energy poverty; modernizing the institutional structures entrusted with managing the energy system; and recognizing and welcoming a wider diversity of actors (technical, financial, entrepreneurial) who can help India navigate these transitions. This task will not be easy. Indeed, the political economy of India's energy system will likely get more complex, with more interests in play. But the same diversity of actors and interests could also induce a healthy, transparent and competitive energy market suited to the diverse needs of India's polity.

Many reports on energy in India focus on specific sectors or sources, and study each in isolation. Yet this approach can lead to technology and policy proposals with unexpected side-effects in separate parts of the energy system. For that reason, our analysis of India's energy options takes a holistic and long-term view and looks at India's energy future, infrastructure, technology, pricing, position in the global energy system and policy environment.

The world that India inhabits now and the world that India will shape through its choices demands that this complex set of filters, transitions and drivers (economic and demographic shifts) be considered carefully. Given the inevitable uncertainties, we have to an extent applied the filters of cost, time and sustainability in a subjective manner to the modelling exercises used in this study. By being explicit about them, we are able to inform the assumptions in our energy models and arrive at value judgments in a transparent manner. At the same time, we fulfil the intention of not being deterministic about the future pathway that India's energy system follows, but drawing on the broad key findings to inform and stimulate a public debate on India's energy choices and the factors that will shape them.

A range of scenarios (using different energy models, which are shaped by varying rates of economic growth and/or technological advances) indicate that India's total primary energy demand could see an increase of around 120% to almost 300% by 2050, an average increase of between 20 million tonnes oil equivalent (mtoe) and 40 mtoe per year. Population growth and economic development will be the two main drivers of energy demand. All projections point to the continuing presence of fossil fuels in the primary energy mix (particularly for use as industrial energy) even as renewable energy enjoys significant growth in five of the seven scenarios developed for this volume. The pace of electrification of the energy system will also determine the rate of build-up of renewable energy infrastructure in India. Attempts to incorporate renewables into the electricity system would also help to diversify the supply mix from an energy security perspective.

Fears may now grow about India's vulnerability, if the contribution of renewables is going to be material but limited. Despite significant increase in renewable-based capacity, fossil-based primary energy could increase from 2 to 4 times its current level by 2050, with imports ranging from 30% to 60% across scenarios. Over the last two decades, energy's share in the import bill has more than doubled, from 15% in 1990 to 38% in 2012, and indeed, by 2050 energy import expenditure as percentage of GDP could range between 5% and 15%.

Therefore, it is clear that India's future energy demand will be large, no matter which projection is considered. There will continue to be a requirement for oil, gas and coal in all these outlooks. And as a result, India must accept that it is going to become an important force in global energy markets. Pushing energy efficiency strongly across sectors is important to contain as much of this energy requirement pressure as possible. Move users from traditional fuels to more efficient and cleaner burning fuels, and providing those who are without energy access to clean energy continues to remain an overriding priority. Technology can play a key role in this transition in a sustainable manner if planned and implemented properly. In addition, energy policy must focus on agriculture, industry and municipal users, keeping in mind specifics associated with these sectors. Finally, India must remain aware and opportunistic in its behaviour as assorted countries push ahead with transitions in their own energy systems, potentially exploiting gains by others to accelerate its journey to a more equitable and resilient destination.

India's energy choices

Coal, gas and renewables in the primary energy mix

With all the challenges that India faces in building a resilient energy system, promoting any single primary fuel over all others, is simplistic. Instead, the role that different fuels must play at different stages of development and across sectors merits further explanation. Conventional options like gas, hydro and nuclear along with coal will need to continue playing an important role in supporting new renewables in the power generation sector. Moreover, with renewables currently having little scope in being able to supply industrial heat, natural gas holds prominence as a cleaner alternative to the use of coal and diesel. Therefore, even as it gives a boost to renewable energy, India cannot veer away from fossil fuels significantly, at least in the short to medium term. In the short term, to provide energy and not hinder economic growth, coal will be the most critical fuel, particularly if the highest priority is providing electricity to all those that require it. But if India is to build an energy infrastructure commensurate with ambitions to limit greenhouse gas emissions, it must seriously consider greater use of natural gas, hydropower and nuclear energy. Natural gas can also provide the heat for a multitude of key industrial processes, but its role in supporting base-load capacity (along with hydropower) could also become essential. Only through building this guaranteed base-load capacity will renewables be fully exploited.

That India should aggressively pursue opportunities in renewable energy is not in doubt. To be stressed, however, is that—for renewables to make a significant contribution—with electricity making up only a fifth of useful energy consumed, any policy on power generation should go hand in hand with initiatives to electrify end-use technology. Even at this point, renewables can only diversify the supply mix, and not become the backbone of the Indian energy system.

For delivering energy to non-urban consumers, recent policies have focused on solar technologies to provide electricity. This is encouraging, but the scale of the residential sector calls for a multipronged approach. India cannot expect to electrify every single household using solar photovoltaics, certainly not in the outlook period, whether through grid-connected or decentralized systems. Yet, with the persistent failure to provide centralized grid-connected power to tens of millions of households, solar PV does have a role to play in partly bridging the current gap in access to electricity.

For cooking energy, India must take into account the role of traditional biomass, now dominant. Numerous technologies now enable the more efficient and safer burning of these fuels. In the near future, then, the focus should be on shifting the population away from burning traditional biomass on open fires and implementing—what is a relatively short-term solution—the use of cleaner cook-stoves and other cleaner burning fuels. This, in turn, buys the time to fully understand the economic and technical challenges of a large-scale rollout for modern cooking fuels (liquefied petroleum gas [LPG] or electricity).

Which transport fuels can India depend on for its growing mobility needs?

All scenarios indicate that between now and 2050 transportation will still rely heavily on oil-based fuels. There may be gradual penetration of electric vehicles, and biofuels can reduce some of the pull on oil products, but at least four of our scenarios have oil products meeting at least 85% of transport fuel demand in 2050.

Of larger relevance is the potential reduction in demand as a result of vehicle efficiency and system efficiency. Energy efficiency could bring significant reductions in energy consumption across various scenarios considered, and ought to be pushed effectively on both technology and policy fronts. The scenarios also indicate that some of the largest reductions can be brought about by reducing the distance to be travelled (through better urban planning) and from modal shifts (shifting more people and freight from roads onto railways, for example). Again, while it is not the mobility-related aspirations that are intended to be curtailed, reduction in useful service-level demands through smart city-level planning and provision of alternatives to personalized modes of movement are required. Similarly, greater uptake of alternative technologies such as CNG or electric vehicles is strongly linked with provisioning of appropriate infrastructure. Enabling modal shifts again is all about infrastructure as well as providing the appropriate nudges to consumers through pricing and policies.

Unlike technological efficiency, where a "rebound" effect sees consumers using more of a product if it is cheaper to run (since it is more efficient), structural reductions in the need for energy service tend not to see the same effects. For that reason, better city planning forms are just as important (if not more important) than policies for end-use energy efficiency. It is estimated that urbanization and the continuing growth of cities could provide an opportunity to lower the need to travel. Urbanization also provides the chance to build a more efficient building stock, which in turn can reduce the energy demand for heating and cooling.

A "no-regret" strategy for infrastructure

India is likely to rely on coal for 55% (or more) of its primary energy needs through the next three decades. All model scenarios suggest that the contribution of renewables to electricity generation by 2030, even with the best efforts, will not exceed 30%. The use of coal could peak by 2040 and the need for assets to mine, handle and transform coal would decline significantly after that, partly driven by the longer term goal to reduce India's contribution to global GHG emissions commensurate with its commitments in the light of its anticipated future income. That said, as elaborated in Chapter 1, when compared against a "business as usual" scenario, India's projected deployment of energy from renewable sources is significant in terms of scale, performance of comparator countries and the timelines within which current policies aspire to achieve large-scale deployment. Moreover, it is likely that India's coal consumption will peak at a lower per capita income than has been committed by China, underscoring how the evolution of India's energy system will internalize the carbon constraint and the imperative of sustainability in setting out its infrastructure choices.

A "no-regret" strategy for the power sector would focus on the rollout of efficient thermal power in the near to medium term and to use it optimally over its technical life. The critical infrastructure components are capacity expansion of the rail network and increasing the output from domestic mines. Model results suggest that transmission of electricity is almost always cheaper than transporting coal. Given the difficulties in acquiring land and expanding the rail network, priority must go to strengthening and expanding the transmission network and to concentrating coal-based generation to areas close to the pithead. Also, given that the country eventually intends to move to cleaner generation options, investments in coal mining and processing infrastructure must also be judiciously planned for. Relying on coal imports may be economically beneficial when compared to the high costs of procuring coal using underground mining techniques and setting up additional new coal washery capacity that may become redundant before the end of their economic lives.

The industrial sector offers significant opportunities for switching from coal to electricity and natural gas—for up to 40% of total industrial energy consumption over the next four decades. This will require creating a national gas grid that would allow industry to access natural gas across the country. Current consumption of natural gas is limited largely to India's western and northern regions. Keeping in mind that natural gas can compete with coal only in the presence of effective implementation of policies to limit GHG emissions, the regulations that govern exploration and production of natural gas from domestic resources could be reviewed to realize the full potential of domestic reserves.

Renewable energy in the power sector must be actively pursued in parallel with other efforts. With increasing contribution to overall generation, grid stabilization assumes importance and needs to be addressed. Flexible sources of generation or electrical storage technologies are not in the portfolio of options available in India today. For this reason, the development of an ancillary services market (frequencies support, voltage control, peaking/operating reserve) must be a priority. A mature market for these services is a prerequisite for the successful integration of renewable-based generation. Notwithstanding the needs of climate change mitigation, a transition to renewable energy is contingent on its price competitiveness with coal and natural gas. The transition will invariably be a two-step process—from coal to gas and then to renewables. An aggressive rollout of renewables will need a large suite of technologies for baseload generation (including gas and hydropower), to help balance intermittent generation and to support mitigation efforts in a meaningful way.

In addition to these "hard" infrastructure choices, there are also some critical elements of "soft" infrastructure, which would apply irrespective of the technologies and designs chosen. The design of legislation is of particular importance and cascades down the value chain. For instance, regulations affecting pipeline construction could have an impact on fuels chosen for new power plants. Further, the government's role will expand but in the form of a facilitator through better regulation (including the autonomy and authority of regulators) and protecting the sanctity of contracts, so that long-term infrastructure investment plans can be made with minimal risk premiums imposed on the costs. Finally, private sector investments in infrastructure could be unleashed by exploring various options, such as instruments to de-risk investments in the energy sector, developing deeper bond markets for energy, and using private sector institutions to upgrade and augment the skills and capacity needed to deliver modern energy services.

Articulating a technology trajectory under uncertainty

Uncertainties in global energy prices and technology development provide the backdrop for India's multiple energy transitions. A sensible energy technology trajectory would therefore need to define clear long-term goals both in end-user sectors and in the energy industry itself, while retaining enough flexibility to move across a band of fuel and technology choices that are competitive and economically attractive. Infrastructure development and policy signals at each juncture therefore need to be geared towards planning ahead in terms of R&D, infrastructure needs and simultaneously avoiding technology lock-in (Box 1).

Box 1.
Key messages for energy technology

Don't expect a silver bullet—India will have to explore all possible technology choices on both the demand and supply sides, investing in a basket of complementary technologies, simultaneously deployed.

Avoid technology lock-in—India plans to increase domestic coal production to 1,000 MT a year by 2021 (in part through high-cost underground mining) and to enhance washery capacity to clean its high ash coal before transport. At the same time, there are plans to scale up solar power to more than 100 GW by 2021. However, at some stage beyond this date, India would have to ask whether it would need both these resources without the risk of potential stranded assets.

Take advantage of being a second mover—There is already significant knowhow, R&D and innovative models adopted across countries. India has adopted compact fluorescent lamp and light emitting diode lighting systems fairly easily, as costs came down sharply due to wide acceptance around the world. The same may be happening to renewable technologies, particularly solar. And clean coal technologies developed elsewhere may be yet another option to which India can leapfrog. Moreover, other developing countries with similar socio-economic contexts may have adopted innovative models and technologies that can be adapted to the Indian context easily.

Come up with decentralized solutions for the hard-to-reach—While LPG penetration may increase, it is not expected to reach 100% of the population. For the unserved, improved cook-stoves are an ideal solution. And solar lanterns and solar home lighting systems could provide lighting solutions in hard-to-reach remote areas without being associated with transmission losses.

The scenarios examined clearly indicate some directional certainties that exist under all circumstances, and constitute robust technology choices for the country. The need for energy-efficient end-use options across each of the demand sectors, a move towards greater electrification of end-use, accelerated and up-scaled integration of new renewables and cleaner power generation options are all near-inevitable. Accordingly, a clear technology trajectory needs to be planned out with timelines of introduction of each technology.

Power technologies

India needs to invest heavily in renewables-based technologies particularly wind and solar. India has the highest potential in these two sources among all the renewables, and tapping them would put India on a track to a more efficient and environmentally sustainable energy system. Since most power projects have a life of at least 20 to 25 years, early investments in these clean technologies would avoid locking in of additional new fossil-based capacities. Under all scenarios that consider a significant scale-up of new renewables, around 150–200 GW renewable capacity is envisaged by 2030 and around 700–1000 GW by 2050.

Even with an ambitious scaling up of new renewable capacities, India's growth path necessitates significant additional generation capacity based on coal, gas, hydro and nuclear. Accordingly, fossil-fuel-based power plants need to be made more efficient. While all new coal-based capacities being set up are already required to be super-critical, an accelerated move to phase out old and inefficient sub-critical plants and move to ultra-supercritical or advanced ultra-supercritical technologies is important. Simultaneously, efficient gas combined cycle power plants should also be seen as a superior alternative to coal in this transition and efforts to enhance the availability of gas for power generation should be part of the focus. Further, recognizing that coal would need to continue playing a role in India's power generation story, R&D on coal gasification technologies and carbon capture and storage must be undertaken to tap this potential if possible.

With the large and growing capacity requirements, the power sector also cannot bank on a single fuel source or technology. Planning for an appropriate mix of fuels and technologies to match up the baseload and peaking requirements in the short, medium and long term with the power demand patterns is one of the key challenges that needs to be addressed. While coal, gas and nuclear would be required to contribute to the baseload requirements, hydro power would also need to play a role in contributing to efficient and sustainable power generation. Nuclear energy has the potential to bring in affordable energy on a large scale, and through a move along the three-stage nuclear programme, India can scale up its nuclear-based capacity significantly.

Industry technologies

In industry, some sub-sectors have seen continuous technological progress, but there is enormous scope to bring current processes to global best practice. In these applications renewables currently provide limited potential, with no renewables alternatives that can currently provide industrial heat above 700 °C, although this should be a priority area for research. Renewables-generated power is increasingly viable as an alternative for light manufacturing, particularly in locations such as industry and technology parks. Accordingly, while R&D for technology in this sector needs to focus on cleaner alternatives, natural gas again could play a larger role to replace coal-based heat and diesel-based captive generation in this sector in the short to medium term.

Transport technology

With rising incomes among large unserved populations, in all scenarios the transport sector (inter-city, rural and within city) is expected to undergo exponential growth. The various scenarios indicate that with aggressive measures the energy savings over the reference scenario could be around 29% by 2030 and 45% by 2050. Prospects for moving away from petroleum-based fuels in this sector are limited, even with the higher penetration of CNG and electric vehicles over time, particularly in urban applications. Biofuels could play an important role, meeting 15% of the sector's energy requirement by 2030 and up to 75% by 2050, if commercial viability of third generation micro algal biofuel could be ensured by 2030.

Per passenger trip or tonne kilometre rail-based transport is significantly more energy-efficient than road transport. The share of rail-based commutes has been falling over the years, and this reduction has to be arrested. Another possible measure is increasing the share of rail in freight (to 50% by 2030) and that in passenger movement (20%) along with enhancing the share of electricity in rail-based freight movement (80% by 2030) and passenger movement (70% in 2050). Introducing mass transport options like metros, improving vehicle efficiency and emission reductions in line with the latest European Union (EU) standards and replacing the old vehicular fleet over time could improve the energy and emission intensities in this sector. In order to bring about a major technology change in the vehicle fleet, the focus needs to be on R&D for third-generation biofuels, hydrogen fuel cell vehicles, bringing in continuous efficiency improvements across all transport modes and encouraging manufacturers to invest in hybrid vehicle assembly lines immediately, so that such vehicles can become the mainstay of the car fleet in the next two decades.

Residential and commercial technologies

With a significant proportion of India's future stock of buildings yet to be built, the main intervention in the residential and commercial sector is energy-efficient building design and the use of efficient appliances, deepening the successful initiatives that are already underway. Energy-efficient building design could be made mandatory, especially in new commercial buildings, while providing a thrust to the manufacture and deployment of efficient (star-rated) appliances, and the accelerated replacement of current lighting systems with LEDs. Focusing research and development on efficient and affordable appliances is important keeping in mind the socio-economic as well as weather-related conditions specific to India. India's Domestic Efficient Lighting Programme, by scaling up public procurement and distribution of LEDs in 100 cities, has slashed unit costs by at least 60%. In the case of household cooking, the transition away from traditional fuels will be gradual. Improved cook-stove technology would therefore remain an important aspect.

Innovating for technology development and deployment

India would do well to encourage an ecosystem for the development of technologies and associated business models in the energy sector. A range of "horizon" technologies could include, on the supply side, solar thermal with energy storage, third-generation biofuels, advanced nuclear, coal gasification, shale gas extraction, and geothermal and wave energy. On the demand side, R&D would be needed in hybrid vehicles, hydrogen fuel cell vehicles, affordable space cooling technologies and high-speed railways, among others. Rapid progress in research and innovation across these elements is required to achieve any significant scale-up or change at the sector level.

Coordinated research that brings together universities, research laboratories and private sector entities and energy technology partnerships with other countries will also be critical. In the past, three obstacles have impeded partnerships: lack of appropriate financing, intellectual property (IP) restrictions and insufficient or under-utilized capacity. Entrepreneurs need upfront financing to cover capital costs of clean energy technologies, working capital to hold inventories and funds to pay IP licence fees. Business models, such as rural micro-grids, could be viable but, being small in scale, they often fail to attract the attention of large institutional investors.

In order to develop new technologies, innovators need clear market signals. Public policy intervention—whether by setting a carbon price, direct R&D investment or guaranteeing a minimum price for emerging technologies—is needed to stimulate private investment. Another approach is to develop partnerships wherein contributing firms/ research institutions retain their original IP but share returns on new technologies. Thirdly, to be inclusive and effective, partnerships should contribute to building capacity. Member countries or institutions need not only contribute in hard currency. In-kind contributions of research staff, facilities or land for demonstration projects could be ways in which the contributions of all members are recognized and duly rewarded.

Innovations would also have to be encouraged for deploying existing technologies adapted to the needs of the poorest 60% of India's energy consumers. More than 400 firms already operate in India in the decentralized energy sector, experimenting with a range of business models. A focus on energy access for consumption and productive uses could increase access to working capital for far-flung small entrepreneurs, pay for licencing fees, underwrite experiments with business models and, by aggregating projects, establish links to larger investors. Initially capitalized by public finance, the revolving fund could attract institutional investment through sovereign guaranteed green bonds. By focusing on deployment, such initiatives could also support skills training in rural communities to build, service and maintain decentralized energy systems, in addition to supporting centres to test and certify products, and create model regulatory codes.

An integrated energy pricing regime sensitive to the needs of India's energy-poor

Energy pricing is complex and politically controversial. This is true in both rich and poor countries. Everywhere, the core challenge is to reconcile economic rationality (also called efficiency pricing) with political acceptability. The former is driven by perceptions of affordability and fairness. The latter requires policy transparency, consistency, credible market and regulatory institutions, political leadership and effective communication.

The challenges of energy pricing (and the related issues of energy taxation) have been widely discussed in the Indian academic and official literature, and are generally well accepted and understood, even if political action has been more cautious.

Encouragingly, the opportunity of sharply lower oil prices has recently been used to introduce greater flexibility in the retail price of diesel, following earlier liberalization of petrol prices. This has improved the financial position of public sector oil marketing companies, given fiscal relief to the exchequer and reopened opportunities for retail competition from outside the state sector. Similarly, a first attempt has also been made to move towards more transparent price discovery for domestic coal following judicial action on the earlier allocation of coal blocks. These are small but important first steps in a much longer journey.

Aligning energy pricing with policy objectives

The focus here is on ways in which energy pricing must adapt to the important transitions facing India's energy sector while still respecting the need to address energy poverty. This requires energy pricing increasingly to be considered in an integrated fashion, both across fuels and along the value chain from production to final user. Since larger public policy goals (fiscal federalism; industrial competitiveness) can sometimes provide conflicting guidance for pricing, it is important to be clear on the core objectives that the energy pricing system should serve, and where other instruments are the better choice to meet equally important social goals.

As discussed above, the key primary fuels for India for the foreseeable future will remain coal, oil and gas, with a growing role for renewables. The first three are all imported at the margin, with India also a significant exporter of oil products. Since imports of crude oil, natural gas and refined product currently take place relatively freely (subject to import duties), the main upstream pricing issues have to do with stimulating domestic exploration and production and ensuring the passing of price signals down the value chain to the end-user (either bulk or retail).

In both these areas, and more widely across the energy pricing arena, it is helpful to distinguish between the price level and the system of price build-up. The price level is the amount a producer receives (or a customer pays) in absolute terms and needs to be set at least to cover costs and to provide a margin, so that the required investments are made to ensure the long-term sustainability of the energy system. The pricing mechanism is the way the price is calculated: either market-based (supply and demand of the commodity), linked to a substitute fuel (such as oil indexation in gas contracts) or regulated on a "rate-of-return" basis. Market-based mechanisms are normally considered the most efficient in delivering the lowest energy cost-effectively to end-users, but they do require a range of well-capitalized players, appropriate infrastructure and effective regulation.

There has been considerable debate on the appropriate pricing and fiscal regime to attract private capital into upstream activities (profit-sharing/production-sharing versus revenue-sharing) but the outcome has been disappointing even when global hydrocarbon prices were buoyant. Current global experience suggests that both fiscal principles can be made to work; that fiscal regimes need to be calibrated to the nature of the asset under consideration; and that transparency and stability are all-important for credibility. In contrast to oil and gas, owing to prevailing legislation, no similar initiative has been taken on liberalizing the infusion of private capital into coal mining.

Looking ahead, it seems that average end-user energy prices will need to remain relatively elevated by international standards, so as to reflect domestic imperatives and circumstances. India will likely remain a major importer of fossil fuels for some decades to come, and needs to encourage a pattern of infrastructure and urban development that is frugal and energy-efficient. This has been the course followed, for example, by both Japan and Western Europe over the last half-century, and is in contrast with the much more energy-intensive trajectory pursued by the more energy-endowed economies of Australia, Canada and the United States.

With growing concern for energy security, prices need to be sufficient to ensure the necessary investment in domestic production and infrastructure. At the same time, competition from imports must remain a credible threat to exert competitive pressure on the domestic energy industry, which implies competitive access to distribution infrastructure. This will take time to establish but ought to be an explicit goal of policy.

The dichotomy of "higher primary energy pricing to attract investment" and "lower pricing of secondary energy to ensure better access" has created major distortions in the energy market in India. It will certainly be argued that "high" end-user prices for energy will undercut both India's anti-poverty and its industrialization ambitions.

Therefore, end-user pricing needs to be focused primarily on the goal of energy conservation and secondarily on ensuring investment in the expansion of supply by both public and private investors. Other instruments, such as direct benefit transfers (for poverty goals), an effective goods and services tax (to limit the effects of tax cascading on manufacturing costs, including exemptions for exports) and a determined effort on the reliability of electric power supply (such that expensive captive power can be reduced, and ultimately eliminated), all form elements of a coherent and holistic strategy that allows these multiple goals to be attained.

Energy subsidies can then sometimes help with managing affordability and accessibility, but they can also create perverse fiscal, macroeconomic, social and environmental consequences. Energy subsidies, especially if poorly targeted, are an inefficient way to provide support to low-income households if rich households capture most of the benefits. The current low energy price environment and the continuing fiscal pressure on the government have opened opportunities for further energy subsidy reform, such as by increasing excise taxes (as an implicit tax on carbon embedded in fossil fuels) or with innovations in direct transfer of cash subsidies for cooking fuel.

A relatively "high" average price for energy is compatible with differential taxation across primary fuels. Countries differ widely in their willingness to use taxation of various fuels in pursuit of social goals, partly reflecting historical tradition as well as geographic conditions. Thus, the more crowded nations of Western Europe have been more willing to impose high retail taxes on transportation fuels to reflect the greater probability of congestion in their societies when compared with, say, the United States. In addition to congestion costs there is increasing awareness of other pollutants associated with liquid fuels in transportation (smog, particulates), which do not apply to other forms of propulsion such as electricity.

Greater volatility is a by-product of the greater linkage with global energy markets which is well-nigh inevitable. While there are obvious challenges of managing the transition, there is little doubt that the overall resilience of the economy is enhanced if price increases are treated as permanent and price reductions as temporary in any administered pricing regime. The ultimate goal should be to aim for as complete pass-through as possible. This of course requires that energy-using sectors have the freedom to adjust their prices which has not so far been the case in sectors such as retail power or fertilizer. The acceptability of a liberalized market for fuels will ultimately depend on faith in the competitiveness of these markets, through some combination of freer entry (what economists call contestability) and well-established independent regulators. These challenges are discussed more fully later in this chapter.

Relying on overseas assets or global energy markets to deal with volatility in global energy prices

Between now and 2030, Indian energy demand is projected to increase faster than that of any other country in the G-20. India's share in daily oil trade then is expected to be 12.5%, up from 7.4% in 2014. It will account for 16.6% of incremental global energy demand by 2035. It will not be the biggest energy consumer, but it will be a "swing voter" in global energy markets with a strong national interest in well-functioning global markets. The needed variety in sources of supply is not guaranteed by solely owning energy assets.

So far, India's response to energy price volatility has been to secure long-term contracts with a few key oil exporters, but that approach is becoming less tenable. For India energy security will not be the same as energy independence.

Such security will require meeting four imperatives: assured supply, safe passage, secure storage and a seat at one or more international forums involved in international energy trade and governance.

While ownership of overseas assets might have a limited role in times of crisis, it has mostly been an ineffective strategy because of low shares of overseas production, a lack of financial resources to compete with other countries, the risks of operating in politically fragile areas and the opportunity cost of not selling energy produced in global markets. Instead, more effort will be needed in boosting India's diplomatic capacity and aligning it with the commercial interests of India's public and private energy companies.

India will need to ensure safe passage of overseas energy supplies. This will be, partly, a function of India's ownership of—or access to—a shipping fleet. Compared with other major energy consumers, India's share of oil and gas tankers is low. Safe passage will also require naval capabilities for India to become a net security provider in the Indian Ocean. India has been pursuing regional as well as bilateral cooperation on maritime security in the Indian Ocean, but such engagements need to be prioritized and intensified. It will also need naval assets that can work with other navies in protecting energy supply routes beyond the Indian Ocean, particularly in the South China Sea, from which new supplies of energy might flow in future.

There is no global energy regime. So in seeking institutions to learn the rules and codes of conduct that govern energy relations between other major economies, India will need to identify the key functions that a regional or plurilateral energy regime could perform and that would otherwise be hard to do unilaterally. These functions include assuring transparency in energy markets, cooperatively managing strategic reserves, jointly patrolling energy supply routes, arbitrating disputes and pooling resources to lower insurance premiums on transporting resources.

As a buffer for emergencies India will need secure storage as well as infrastructure and management capability to store and transport energy resources, within and outside its territory. Strategic petroleum reserves capacity of 5.33 MMT (about 40 million barrels) has been developed at three locations, to be commissioned and filled by the end of 2015–2016. Additional capacity of 12.5 MMT (93 million barrels) is to be commissioned by 2020. Storage capacity is low, however, relative to that in OECD countries or in China (current and planned)—and capacity alone would not suffice in any case.

India can learn here from the experience of members of the International Energy Agency (IEA) in two key areas: siting storage facilities and developing the institutional capacity to manage strategic reserves. Any new construction entails heavy financial commitments ($615 million for just building storage for 13 days' worth of oil). An alternative route to building storage facilities is to site some emergency crude oil stocks in three or four other countries, as is the practice among IEA members.

India has drawn on the OECD's three-stage mechanism of establishing oil emergency response organizations, stockholding oil, and implementing oil-stock drawdowns and other emergency response measures. But it needs a regulatory body to oversee the entire strategic petroleum reserve (SPR) process. It needs to coordinate with the government, the oil marketing companies and India Strategic Petroleum Reserves Limited to ensure concerted action—from building and filling up the SPR to releasing crude oil and petroleum products in event of emergency. And it needs to replenish and maintain the SPR levels once normalcy has been restored. It is also important to have storage facilities for crude oil as well as petroleum products, for which an Oil Stockholding Act could be considered.

An integrated policy environment

Given the complexity of the energy system and the federal structure of the government, it is imperative to coordinate policy across sectors. This coordination must be extended beyond energy supply and demand sectors to the institutional structures. An integrated energy policy can work only if it has an enabling environment and institutional structure ensuring efficiency and fairness.

Industry. Given the negative externalities that coal-based power produces and the imperative to move towards renewables, a dynamic policy should incorporate the electricity requirements of industry. The government should provide a consistent policy to enable industry to move away from captive power generation. Developing a smart grid that enables a two-way interaction and increases reliability is essential for industry.

Similar action is required in planning for oil and gas demand. The fertilizer sector has shown a strong move away from naphtha and towards natural gas. It is expected that the demand for fertilizers will increase threefold by 2050 and, given the paucity of natural gas, it is important to consider how this demand would be met, especially in a politically sensitive and relatively inflexible sector like agriculture.

The government has shown a strong commitment towards combating climate change by monitoring energy efficiency through the Perform Achieve Trade (PAT) scheme under the National Mission on Enhanced Energy Efficiency, an example of a well-planned and consistent policy. The policy rollout has been a success, and as the performance of the first phase (2012-2015) gets audited, the need now is to maintain the momentum and include more sectors within the ambit of the programme.

Transport. The transport sector has seen an expansion in personal vehicles, with hydrocarbons as the main fuel choice. It needs policy support to enable both a modal shift to public transport and a smooth and gradual transition towards greater use of biofuels and electric vehicles. The National Electricity Mobility Mission is an important step in promoting the uptake of electric vehicles by providing fiscal support. But it is also important to incorporate the allied infrastructure, such as charging stations required for successful implementation of electric vehicles.

A forward-looking biofuel policy is also important. Detailed policy guidelines have been provided for the first generation biofuels—from non-edible oil crops and bioethanol from sugarcane molasses. But the scope of expansion is hampered by limited feedstock and large land requirements. More advanced biofuels (lignocellulosics) have been identified as potential fuel options and there has been mention of encouraging R&D and financial support. But the current policy is only aspirational, and a clear roadmap is needed. Promising future biofuel feed stocks relevant to India must be identified. Gaps in conversion technologies need to be spelled out. An intensive national biofuel R&D programme, which creates awareness, encourages extensive research and has clearly defined goals, should be taken up.

Agriculture. The agriculture sector has been growing at 4% a year since 2000. To boost its growth, measures will be required to improve irrigation and facilitate mechanization, and this would increase the energy requirements for land preparation and irrigation. The National Mission on Agriculture Mechanization envisages increasing the penetration of tractors and tillers. And the inter-linkages with water and other resources have prompted the government to introduce policies for electricity pricing, micro-irrigation, crop diversification, and energy efficiency. The sheer number of overlapping policies creates challenges of communication and coordination to ensure that the sector's growth and modernization are not compromised.

Residential and commercial. It is estimated that 70% of the buildings required in 2030 are yet to be built. A forward-looking policy can prevent locking in energy-inefficient setups. Moving towards a mandatory Energy Conservation and Building Code should be considered. And since much of the urban infrastructure is yet to be put in place, proper urban planning and development of smart cities should be taken up now. In the current framework for smart cities, the focus is on walking, cycling and public transport. Growing vertically in smart cities is essential, since horizontally developed cities require travelling longer distances.

Only 30% of the Indian population lives in cities, a number expected to rise to 38% in 2050. It will be important to simultaneously increase living standards in rural areas and prepare for smart villages there along the lines of smart cities, possibly through decentralized options to enable the transformation to a resilient energy system.

The commercial sector in India is also growing at a fast pace, owing to lifestyle changes and higher incomes. Shopping malls, hospitals and large office buildings have very high energy consumption, and many have been constructed in water- and resource-stressed locations—another area for urban and holistic planning. Appropriate energy pricing and regulatory signals could encourage the commercial sector to adopt more energy-efficient buildings, appropriate resource mapping and planning, and incorporating decentralized energy systems to complement centralized infrastructure.

Institutional linkages. The institutional structure in the energy sector in India is quite complex, with several institutions at the state and central level and overlapping areas of jurisdiction and authority. This complexity, while natural to a federal structure, often impedes effectiveness. For example, some of the public sector generation companies are under the central government, while the distribution companies (DISCOMs) are under various state governments. Therefore, the generation companies would pass on an increase in fuel prices to the DISCOMs, but the DISCOMs are often unable to pass the higher prices on to the final consumers due to political pressures from state governments. This leads to an inefficient structure where DISCOMs bear the losses, which are in turn passed onto the exchequer.

In sum, India needs an integrated energy policy that is stable and consistent and has a long-term vision. Such a policy would address the linkages between demand and supply for energy and non-energy sectors. It would ensure consistent supply of resources and map the intermittence or pattern of availability of the resource and incorporate this in the planning process. It should also manage and enable institutional structures required for a smooth transition to more liberal markets. Only an integrated policy environment can assess the needs across sectors and institutional structures. It would juxtapose often opposing challenges and interests and offer clearer direction to the stakeholders. And it would build trust in the system and ensure that it is possible to implement tough decisions, thereby reconciling economic rationality with political feasibility.

CHAPTER 1

India's Energy Future

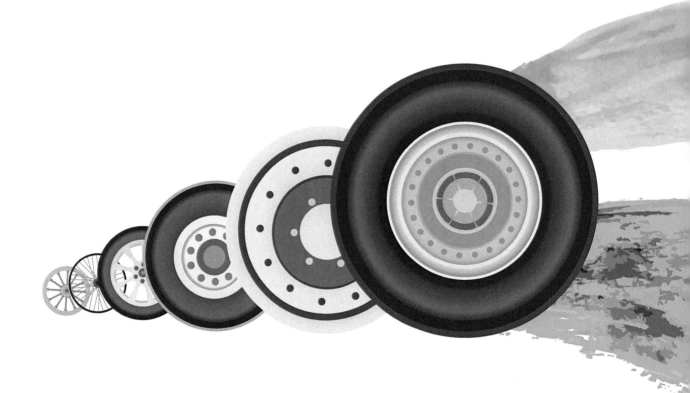

By 2050 India may become the world's second-largest energy consumer, behind China. Its decisions today will determine the shape and success of its energy system and have a major bearing on global energy and emissions.

India is aspiring to expand its economy and achieve development goals that require reliable, basic and clean forms of energy for all its inhabitants and industries. It must provide this energy by building infrastructure and markets that give it energy resilience in an increasingly volatile world. Marrying these two objectives is challenging yet achievable. It requires acquiring and adopting the right technologies at appropriate times and scales. It also requires planning ahead for supporting infrastructure to seamlessly integrate alternative fuels and technologies into the system, and having in place appropriate policy and regulatory regimes that send clear signals to all stakeholders. And rather than shy away from becoming exposed to global energy markets, it should learn to play the game that other major energy importers and consumers have played so well in the past.

Four elements must be optimized to create a well-balanced energy system: infrastructure, technology, pricing and integration into global energy markets, all bound together by policy and political rationales.

India is home to around 1.25 billion people,[1] a population expected to grow to about 1.75 billion by 2050.[2] And in working towards its goal of eradicating poverty and providing basic infrastructure, services and clean energy access to all, the Government of India aspires to rapid and inclusive GDP growth that could double per capita incomes every decade. Against this setting, energy projections from all studies indicate that energy demand would increase, doubling or even tripling by around mid-century.

Four main transitions are under way: from traditional fuels to modern; from rural to urban; from energy island to full integration into global energy markets; and from conventional fossil fuel to low carbon technologies to ensure environmental sustainability. India's policymaking has so far been geared towards providing everyone with access to energy, attempting to reduce fuel import dependencies and limiting carbon emissions without stunting the nation's growth. This book argues that these goals are too simple. It also argues that high import dependence should not be viewed with apprehension.

Many other countries have prospered while being net importers of energy, all because they accepted and embraced their role in global energy markets. India until now has not had to worry unduly about global energy markets—as a relative price taker. But it could gain more by increasing its engagement in global markets—perhaps to the point of even becoming a price maker. Accordingly, policies should be reframed to focus on ensuring the availability and affordability of appropriate energy fuels and technologies to different sectors and sections of society, embracing global markets and undertaking a more integrated and holistic resource planning approach over the long term.

The first transition is from traditional to modern energy fuels. Incomes in India are rising, yet 21% of the population lives below the poverty line. Nearly two-thirds of Indian households continue to use firewood, dung-cake and charcoal for their cooking needs. Rising wealth will see the second transition from traditional to cleaner modern fuels such as electricity and liquefied petroleum gas (LPG). But there are issues of both the availability and the affordability of these options to those in the lowest income brackets. While people indicate a willingness to pay a premium for reliable and clean energy, fuel stacking (traditional fuels supplementing modern fuels in rural households) is clearly observed in many low-income households that have access to clean energy. In the interim, multiple technological solutions such as improved cook-stoves may for some time remain important solutions to India's clean cooking transition—before gaseous fuels or electricity become widely available.

While the total population is expected to increase by nearly 50% by mid-century, the urban share is forecast to reach 38% (from 31% in 2011)—665 million people in a country of 1.75 billion. A growing proportion will be aspirational middle classes with rising energy demands. Cities, if planned and managed well, can reduce the footprint of energy, water and land that are used and required by the population. Managed badly, they only make these resource stresses worse. This is the second transition.

Third, India will move from an energy island to a fully integrated major global player. Even with fairly significant energy efficiency assumptions in energy demand estimates, by 2030 India's total primary energy demand is projected across various studies and scenarios to increase between 1.9 and 2.7 times the 2011 level, and between 2.2 and 5.3 times by 2050/2051. Moreover, with the current substitution possibilities across end-use sectors, and the limits to immediate scale-ups, all projections point to the continuing presence of fossil fuels in the primary energy mix. India should not be afraid of high import dependence, but it should intelligently evaluate the benefits and trade-offs from investing in domestic production versus imports over the next few decades in planning its fuel and technology transitions. Whatever the outcome of future talks on climate, more and more countries are decarbonizing their energy systems, due to various other co-benefits associated with clean and efficient alternatives. India's National Action Plan on Climate Change has clearly delineated energy efficiency, enhanced electrification of the energy system, a significant thrust to renewables and sustainable habitats as some of the key elements of India's decarbonization story. More recently, India's Intended Nationally Determined Contributions (INDCs) again reiterated pursuing energy efficiency and renewable-based development as a key element of India's way forward in the energy sector. There is an increasing realization among government and private actors that, with India's future supply and demand projections, they will be operating in a much different energy environment. And with the world's energy systems in transition, India has an opportunity to take advantage of new decarbonization technologies along the route to a sustainable national system. This is the fourth transition.

The analysis here, when broken down into its components, points to India being on the right track directionally in developing its energy system. India plans to move towards increasing electrification, enhancing non-fossil energy, increasing more efficient rail-based movement and gradually adopting cleaner fuels and technologies. The capacity to undertake larger transformations is, however, likely to be limited by the required supporting infrastructure, finance and skill-sets. Changes on both the energy demand and supply sides are likely to be incremental and spread across multiple energy sources and technologies—both existing and emerging. Moving to a scale and pace of change that is desirable would require alternative fuels and technologies to be available for adoption and up-scaling much more rapidly. It is against this background that the pursuit of R&D and innovation assume relevance for technological solutions and financing models suited to India's context.

Coordinated attention is needed in four areas:

- Creating a step change in reducing energy poverty among India's citizens, boosting energy equity.
- Attracting investment from diverse sources for its future infrastructure needs and avoiding the risk of stranded assets.
- Determining how India becomes more deeply engaged with global energy markets.
- Moving towards more modern institutional structures (including those of international engagement) to create an integrated energy system that works for all, ensuring sustainability in India's energy situation today—and through mid-century.

Population growth and economic development are the two main drivers of energy demand. Between 2001 and 2011 India's population grew from around 1 billion to 1.2 billion, with economic growth averaging 8% a year. Total primary energy demand grew at 5% a year; in 2014, 70% of it was met through fossil fuels (Figure 1.1). Coal and petroleum were the main fuel supply sources at 39% and 23%, respectively, while natural gas contributed 8%.[3] Residential, industrial and transport were the main consumers in 2014. In total final energy consumed, India that year used 20 exajoules (EJ), or around 478 mtoe.

So, how does India compare with other countries previously at its income on the energy ladder? As of 2012 it is on track, with energy intensity as predicted for its income (Figure 1.2 and Box 1.1). How its energy mix would play out over the next decades, however, is still uncertain. North American countries, particularly owing to their sprawling land use (and cheap energy), are very energy-intensive. It is not likely that India will end up at these levels—nor should it aspire to. Western European countries, in contrast, see lower intensity due to higher prices and greater energy efficiency (partly due to more imported forms of energy). So the energy implications of India's future development trajectory would depend on the progress of technological development and diffusion globally, the availability of finance, the status of domestic policy reforms and regulatory structures, and its role in global energy markets.

Figure 1.1.

Fossil fuels met 70% of India's primary energy demand in 2014

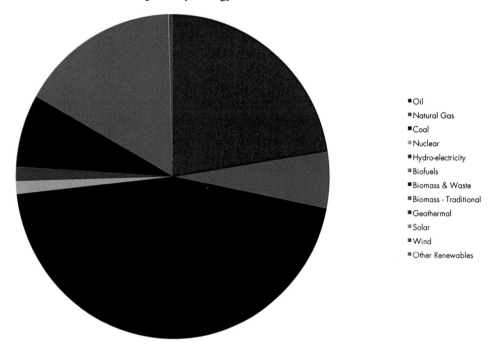

- Oil
- Natural Gas
- Coal
- Nuclear
- Hydro-electricity
- Biofuels
- Biomass & Waste
- Biomass - Traditional
- Geothermal
- Solar
- Wind
- Other Renewables

Figure 1.2.

India is on the appropriate rung of the energy ladder, 1960–2012

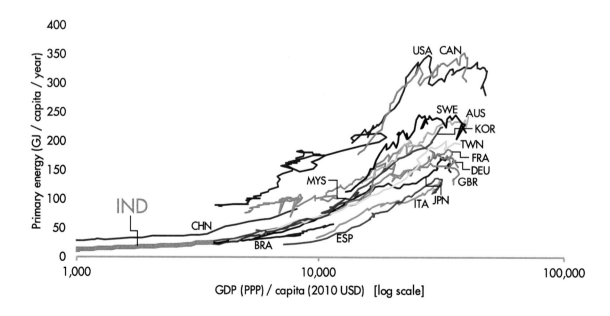

Box 1.1.

Energy demand grows with income

Economic growth and energy demand show an inexorable relationship—as a country gets richer, it consumes more energy. But the relationship is not linear—it tends to follow an S-curve. As the very poorest countries start to get richer, they tend not to use significantly more energy. When average incomes reach $4,000 a person (in purchasing power parity, or PPP), demand tends to accelerate, usually as a country enters its stage of industrialization. In other words, energy intensity rises. At around $15,000 a person, demand growth eases, as some uses approach saturation (for example, steel production) and the economy diversifies from industrial to service activity. But even at the highest incomes observed today, the total energy consumption per person still rises.

While there is consistency in the overall pattern, there is some variation in the exact acceleration point. China started accelerating at the typical $4,000 a person, a point it passed a few years ago. Thailand and Malaysia began accelerating at closer to $3,000 a person, and Brazil not until around $6,000 a person.

The point when energy demand growth slows varies significantly. Countries with higher primary energy demand (250–350 GJ a person a year) tend to be correlated with more dispersed land use (United States, Australia), be in colder regions (Canada, Finland) or have abundant energy resources (United States, Russia, Gulf countries). The EU and Japan, with higher population densities and lower indigenous energy resources, have primary energy demand of 150–200 GJ a person.

After income, price is the second most important factor determining long-run energy demand in a country. Examining price effects tends to help understand some—though not all—of the differences between country paths up the energy ladder. Major resource holders have cheaper energy available and have often subsidized its use. Countries often began their acceleration during periods of relatively low energy prices (Thailand 1980s, Philippines 1990s). Assuming a typically European or American development pathway for developing countries makes an enormous difference to any outlook for the world's energy demand in 2050.

Developing countries are likely to follow pathways towards the lower half of the range (Figure 1.2 in the main text) for three reasons. First, there is a (weak) tendency for countries that have developed in more recent times to have lower energy ladder pathways than countries that developed decades ago. Second, the econometric equations to extrapolate developing country demand based on the derived elasticities leads to pathways similar to those in the EU and Japan today. Third, extensive and wasteful use of energy is less likely than in the past.

Source: Shell's World Energy Model.

With a need to balance rapidly increasing demand with reliable and clean energy supply as much as possible, some may argue for India to follow the pattern of Southern Europe, where countries show some of the lowest intensity among developed economies. In contrast, both South Korea and Taiwan are resource-poor economies which chose economic pathways that relied on the creation and growth of heavy industry. Now at GDP per capita levels on par with the world's richest countries, they are beginning their transition into economies with more services and personal consumption. And both took advantage of low global energy prices to begin industrializing. The deregulation of prices was also crucial in facilitating the modernization of those two economies, and India would be taking a risk if it chose to adopt this approach, not knowing how long today's low prices might last.

However, each country is unique in terms of its resource endowments, its economic structure, population distribution, geographical spread and features, making it impractical to emulate any other country's development pathway.

Large countries like India tend to spread out more, increasing the pressure on transportation services. Moreover, energy and emissions may differ simply in the way we account for consumptive versus territorial emissions, that whether extraction and conversion of energy and production of goods occurs within the country itself or happens outside the geographical boundaries.

Structurally, countries with higher-value-adding industries are likely to have lower emissions intensities than those with a large share of energy-intensive manufacturing and a large labour force that needs employment.

To explore how India's energy system pathways could develop, we considered selected outlooks from The Energy & Resources Institute (TERI), the Council on Energy, Environment and Water (CEEW) and Shell's Scenarios Group. Details of the models and the results are in Appendix 1, and the high-level assumptions in Table 1.1. The key takeaways from the modelling scenarios are discussed throughout this chapter.

Table 1.1.

High-level assumptions of seven scenarios

Institution	Scenario	Assumptions
CEEW	Watch and wait (W&W)	Developments in the energy system progress as with current policies, albeit with an awareness of the impact of unbridled energy consumption. No constraints are imposed on economy-wide CO_2 emissions. Interest is sustained in furthering the role of coal and the rollout of renewables to the extent that it is needed to bridge the gap that fossil fuel-based sources leave.
CEEW	Low-carbon inclusive growth (LCIG)	Industrial, residential and commercial sectors exhibit high levels of efficiency. A switch to efficient fuels is encouraged in industry. There is a significant shift in the power sector mix where aggressive rollout of renewable sources is a policy target.
Shell	Mountains (MTNS)	Slower economic growth, and rigid institutions and structures slowing reform but more successful implementation of policy and top-down initiatives. Coal remains the backbone of the energy system yet natural gas grows in importance.
Shell	Oceans (OCNS)	Significant reform driving faster economic growth and greater empowerment of the populace runs counter to many proposed policies. Distributed and localized schemes are the most effective way of driving infrastructure and consensus. Large breakthrough for solar PV, though reliance on oil remains.
TERI	Business as usual (BAU)	Official economic and population growth rates, no major policy or other interventions assumed, current trajectories assumed to continue with little stresses on supply or demand.
TERI	High renewables (RES)	Aims for reduced imports and increased dependence on domestic production on the supply side—and for efficiency improvements and technology leap frogging on the demand side. Includes the added angle of environmental conscientiousness, moving heavily towards renewables.
TERI	Increased domestic production (UCG)	Also aims for reduced imports, though in this case by featuring higher domestic production of fossil fuels, including unconventional sources like coalbed methane and shale gas and less stringent environmental policies.

How might India's total primary energy mix look in 2050? While there is some variation, the continuing role of fossil fuels is clear whichever scenario is studied, but the magnitude of renewables penetration clearly is sensitive to the scenario (Figure 1.3).

Figure 1.3.

Renewables penetration is clearly sensitive to the scenarios

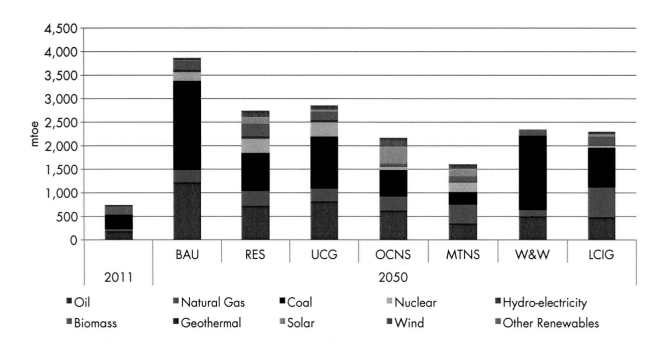

On an energy-intensive industrialization pathway

India is yet to set up basic infrastructure for providing education, health services, road connectivity and housing to all its people as it develops, so the reliance on energy-intensive industry is expected to continue. Given the large amounts of energy to produce iron, steel, paper, cement and aluminium—and their relatively low contribution to economic value-added for the scale of their energy use—India's energy intensity could rocket (until its economy shifts to light manufacturing and services, likely around 2040–2050). More than requiring vast amounts of energy, this initial stage will require large amounts of high-quality heat, which for the foreseeable future will have to be delivered by fossil fuels (Figure 1.4) and their derivatives because electricity cannot deliver the "correct" form of energy (high-quality heat/steam) (Box 1.2).

Box 1.2.

Why fossil fuels are needed for an industrializing economy

Fossil fuels are essential in most fundamental industrial processes, such as chemical manufacturing. The ubiquitous mobile phone and the much talked about solar PV cell are just the tip of a vast energy-consuming industrial system, built on base chemicals such as chlorine, but also making products with steel, aluminium, nickel, chromium, glass and plastics (to name but a few). The production of these materials alone exceeds 2 billion tonnes annually. All of this is of course made in facilities with concrete foundations, using some of the 3.4 billion tonnes of cement produced annually. The global industry for plastics is rooted in the oil and gas industry as well, with the big six plastics all starting their lives in refineries that do things like convert naphtha from crude oil to ethylene.

All these processes are also energy intensive, requiring utility scale electricity generation, high temperature furnaces, large quantities of high pressure steam and so on. The raw materials for much of this come from remote mines, another facet of modern life that OECD nations no longer see.

Economic growth requires consumption of huge amounts of raw materials, all of which require energy-intensive processes to convert them into useful products and goods for society. These large-scale processes will inevitably draw on the energy-dense fossil fuels, since in many sectors electricity is no substitute.

Germany already has an existing and fully functioning fossil fuel and nuclear baseload generation system installed, which can easily take up the slack caused by intermittent renewable-based generation. But the cost is almost never included in the assessment of the cost of renewable power generation. In Germany this is a legacy system so it is taken for granted. But for countries now building new capacity and extending the grid to regions that previously had nothing, this is a real cost that must be considered.

The German experience shows that you can shift to renewables more easily when you already have a fully depreciated fossil and nuclear stock and your demand is flat. Otherwise, this could be a potentially costly story that relies on storage technologies that are not yet in mainstream commercial use.

Source: Adapted from http://blogs.shell.com/climatechange/2014/03/heartland/

Figure 1.4.

Fossil fuels to supply majority of industry's energy needs, even in 2050

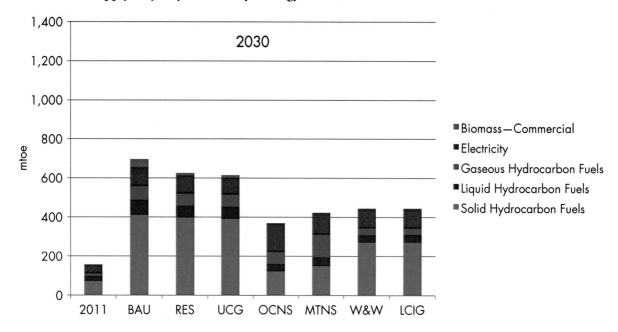

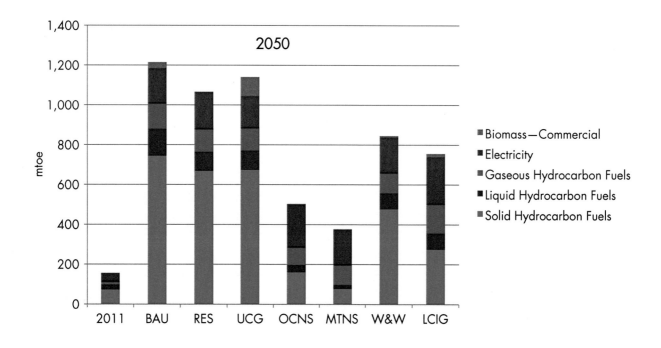

Climbing the energy ladder

All the seven scenarios indicate that India's total energy consumption would increase significantly by 2050. A significant level of efficiency improvement across sectors is already internalized into all the scenarios (to reflect stringent assumptions on global improvements in energy efficiency as well as effective policy in India), ensuring that the country remains on a more efficient lower rung (Figure 1.5). Since India's capital stock is (and will remain) relatively young, it has the second mover advantage in many cases and may have several choices to incorporate improvements in energy efficiency. However, technologies as they exist globally may need innovation and adaptation to India-specific socio-economic and climatic factors to be successfully integrated. Further, effective pricing and policy measures to contain the rebound effect would be important.

Figure 1.5.

Still on the lower rungs in 2030 and 2050

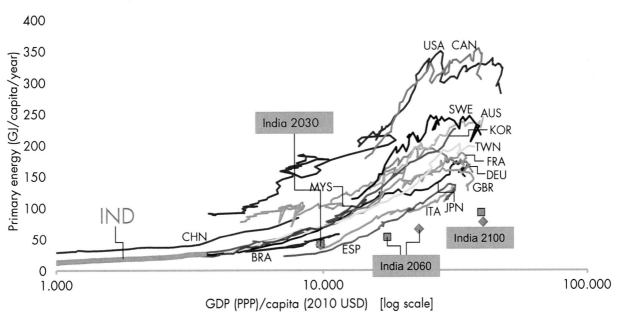

Source: History—IEA; Outlook—Shell Scenarios.

Increased role for renewables

With industrialization causing a surge in demand and fossil fuels evidently irreplaceable in some areas, are there ways for renewables to feed India's industrial growth aspirations? Most countries' consumption is split into three main areas: industry, residential and commercial, and transport. For India in 2011, industry alone accounted for 40%, at nearly 190 mtoe (Figure 1.6).

Figure 1.6.

Industry took 40% of final consumption in 2014

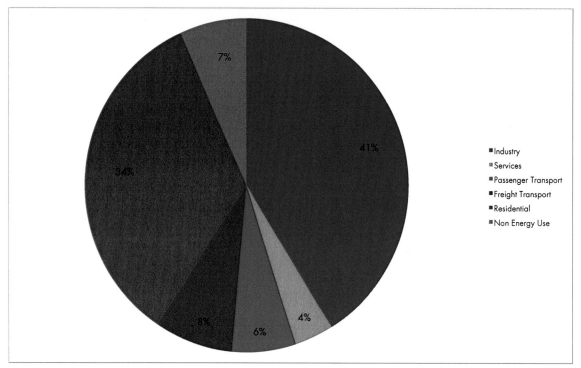

Source: IEA.

Renewables-based generation, more expensive than coal, is further complicated by its intermittent nature and a grid that is not yet ready to manage variable loads and ensure smooth power evacuation. Before committing itself (and the associated capital) to a fossil-heavy future, should India instead consider shifting large sums of money to a more expensive, yet greener renewable-based system? And how far should a resilient energy system rely on one primary energy source supplying a disproportionate share of the mix, which would affect system stability and security and distort the regulatory climate for public and private investors?

First, renewables largely generate electricity (the discussion here focuses on the two main sources, solar and wind). This may seem obvious, but its significance can be demonstrated with a basic understanding of the energy system. Primary energy sources cannot really be consumed in their raw form. Primary energy sources (oil, gas, coal) must be transformed into energy carriers (electricity, petrol, diesel) so that the end-user (residential, industry, transport) can actually use the energy (Figure 1.7). However, some energy is lost as a primary source is transformed into an energy carrier (averaging 40–50% for electricity, and 10% for petrol/diesel).

Figure 1.7.
From primary energy to carriers and to final use

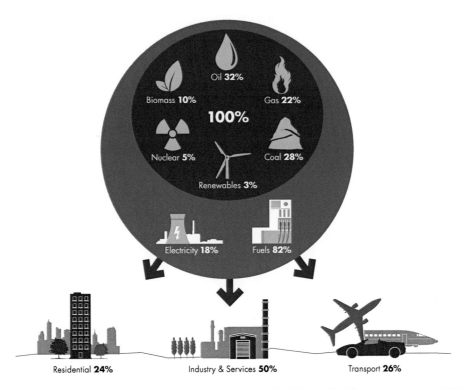

World Energy Demand, 2012. Total primary energy = 556 EJ/year. Total final consumption = 373 EJ/year
Fuels (82%) comprise fossil fuels (67%), biomass (12%) and commercial heat (3%)
Sources: IEA Extended Energy Balances 2012 and Shell Scenarios Team 2013.

So, while electricity is an energy carrier used across all the end-use sectors, it makes up only a fifth of India's total final consumption today. The balance comes from liquid/gaseous/solid hydrocarbon fuels, along with various forms of biomass. The transport sector in particular is largely dependent on petroleum fuels accounting for only 2% electricity, while industry has 44% of the demand based on electricity, and residential sectors 22%. Any drive to increase the use of solar and wind sources would thus need to simultaneously enhance the rate of electrification across each of the end-use sectors. This brings several other challenges associated with the preferences of consumers, the amenability of electrification to end-uses and the costs of infrastructure associated with electrification. Consider the potential investment and rate of build required to bring in large amounts of renewables in the extremely aggressive case of Oceans—up 350% from business as usual (Figure 1.8). Even in Mountains the investment compared with Business as Usual is around 175% greater than 2010 levels.

Figure 1.8.

Electricity fuel mix in generation 2030 and 2050

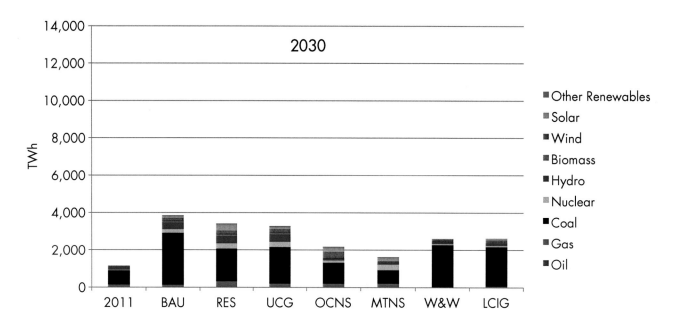

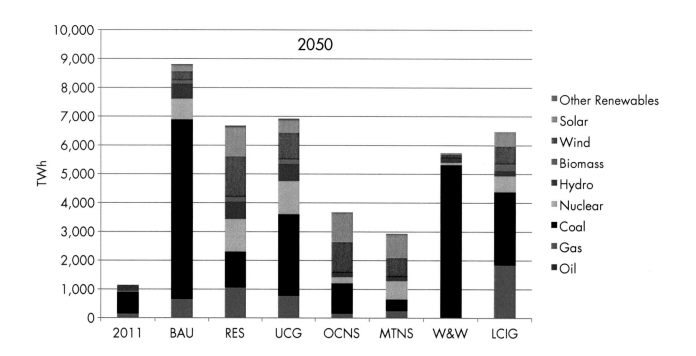

Even in Oceans, more than 60% of primary energy would still be fossil-based. The estimates from CEEW's India National Energy Model (INEM) also indicate that, even with determined goals to cut CO_2 emissions (LCIG scenario), more than 85% of the total primary energy is likely to come from fossil fuel-based sources. Some might argue for an even more aggressive roll-out, but India is performing at similar levels to China in its demand for renewable energy at varying levels of GDP as per Shell's New Lens Scenarios (Figure 1.9).

Figure 1.9.

On renewables deployment India matches China in Mountains and the OECD in Oceans

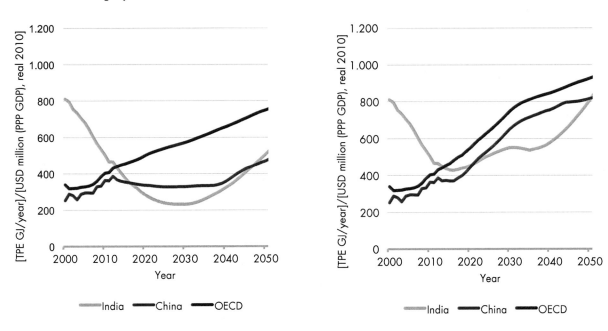

India is already on the right path, by indicating, in its submission to the UN Framework Convention on Climate Change (UNFCCC), an aspiration to deploy renewables aggressively so that non-fossil sources account for 40% of India's electricity generation capacity by 2030.[4] Plausible alternative future pathways from the scenarios here show that solar and wind can both grow to supply a useful share of total energy demand. But this must be tempered with realism. On cost, solar may out-compete fossil by the mid-2020s on the levelized cost of electricity, but system costs will also add to the bill. So, a very large share of renewables should be adopted only once these technologies are economically more viable and their applications technically sound. It may be prudent to focus initially on niche sectors and user groups that could benefit from renewable applications. In the interim the focus could be on R&D to bring costs down further, improve efficiencies and align supporting frameworks and infrastructure to ramp up capacities. The prime candidates include remote rural areas that are not connected to the grid or urban commercial centres (hotels, hospitals, office buildings) where poor quality grid-based electricity is responsible for near complete dependence on captive generation using expensive and polluting diesel.

For India, intermittency remains an unaddressed challenge. Using the latest data, Shell analysis shows that in Germany, the intermittency from large installations of wind and solar PV can be managed using existing fossil fuel and nuclear baseload generation.[5] Renewable advocates may omit the cost of this back-up generation in their arguments for rapidly deploying technologies. For a country with a well-developed legacy energy system, these base-load providers are taken as a given. But when designing new capacity, planners and other analysts cannot neglect the generation capacity required as a baseload to counter renewable's intermittency, and that demands a huge investment. CEEW research finds that the planned 100 GW of solar capacity would require an additional investment of $20–$33 billion in gas-based back-up capacity, over and above the $100–$113 billion of capital investment in the solar projects alone.[6]

Renewables could be very important in providing access to electricity in remote areas that are not connected to the grid. But the rapid scale-up of renewables is likely only with better storage technology, the availability of a supporting grid and conducive policies to enable renewable uptake. Again, India needs to consider a judicious plan for the transition to renewables—carefully analyzing the tradeoffs related to locking in fossil-based options while fast-tracking renewable-based generation capacities and its support infrastructure. With energy shortages of 5% and peak shortages of 2% in 2013, the country cannot meet its energy demands today. And with envisaged GDP growth of more than 8% a year for the next couple of decades, India would have to continue to rely on fossil-based capacity generation as well.

Importing fossil fuels: essential for an industrializing India

For many years, global emission regulation conferences have emphasized the need to modernize energy systems by shifting from carbon-heavy fuels to cleaner alternatives. But a continuing surge in upstream oil and gas investment, a huge rebound in North American oil and gas production, and concerns about the burden of investing in renewables at an early stage of a country's development temper the urgency of that shift.

In little more than a decade the proportion of India's fossil energy that is imported more than doubled, from 15% in 1990 to 38% in 2012. By 2050 fossil energy import dependency could range between 40% and 90% across the energy scenarios (Figure 1.10).

Figure 1.10.

Energy imports could range from 40% to 90% by mid-century

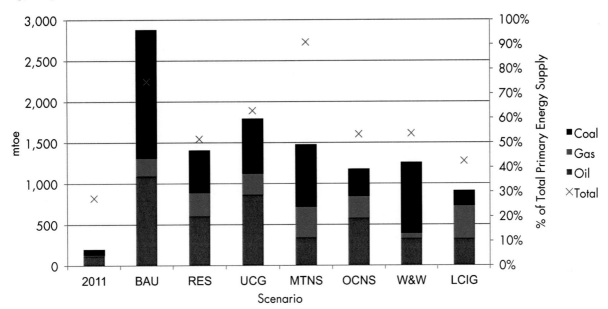

Some may argue that such volumes do not take into account the inevitable efficiency gains, but the opposite is true. The Oceans, High Renewables and LCIG scenarios have very aggressive assumptions for efficiency, so even though policy- and price-driven initiatives will help keep volumes down, they cannot completely remove the need for continued imports. And those who argue that unleashing India's shale gas potential will have a substantial impact on trade balances will see that, while slightly reducing gas import volumes, it does not materially alter the picture.

That said, projected import costs may trigger the question of "energy security" in many people's minds. But over the last few years, informed deliberations have steadily moved away from focusing on energy self-sufficiency alone to including elements of the adequacy, affordability, reliability and overall sustainability of resources in the long run (Chapter 5). Understanding what energy security is and how it should be defined is still not quite settled. But energy security is not the simple aim of minimizing the share of imported energy in the primary energy mix. It is more an issue of guaranteeing secure supplies to final users of energy and reducing the threat of geostrategic or price shocks emanating from abroad. Energy security is fundamentally about ensuring that energy disruptions do not interfere with economic activity.

Consider India's position today. Despite having one of the largest coal resources in the world, India's domestic production has not been able to keep pace with its growing demand for coal. From importing a mere 23 MT in FY2002–2003, it imported more than 140 MT in 2012–2013. Despite a decline in recent years, across the scenarios the results suggest that India may require coal imports of some 740 to 2,140 MT in 2050. For oil, imports could rise from 2 million barrels a day to 6 or 7 million by 2030. For gas, imports could be around 100–150 mtoe a year by then.

These numbers might seem high. But how do they compare with other large energy-importing nations such as Japan? For oil, despite India's possibly becoming the world's largest oil importer, its import share as a proportion of GDP would be no more than around 2%—Japan's in 2010 was 3.6%. For gas, Japan's imports were 0.9% of GDP in 2010—a share that India will be unlikely to surpass even by 2050, according to the Mountains and Oceans projections. Through a balance-of-payments lens, these future import volumes are manageable. So, if India wants to meet the needs and aspirations of its people, as of now it simply does not have the option of using alternatives to oil at a sufficient scale.

Transportation for a growing population

Another important demand segment is for transportation, predicted to grow rapidly as the population and incomes rise. When comparing outlooks across the road transport sector, even with aggressive assumptions for the penetration of biofuels and energy efficiency, the sector will continue to rely on oil products for the majority of its needs (Figure 1.11). This points clearly to a lack of adequate fuel and technology alternatives. Energy efficiency also shows up as a key element to contain the spiralling demand for petroleum fuels in the shorter term.

Figure 1.11.

Transport will continue to rely on oil products for most of its needs, 2030 and 2050

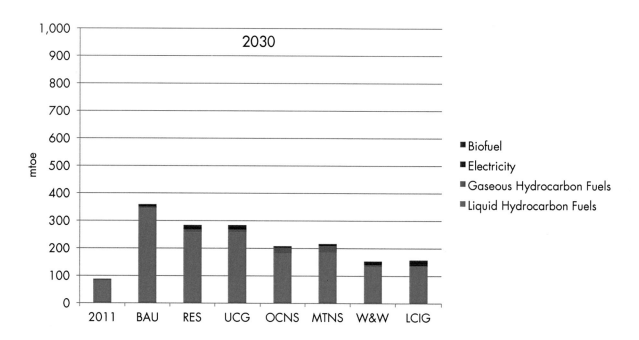

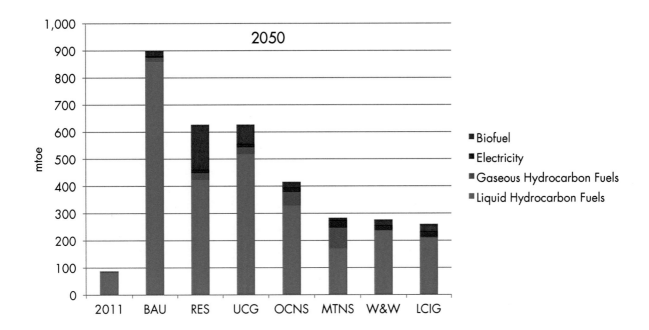

India should pay particular attention to its imports of natural gas, since global LNG volumes are expected to increase as the LNG business continues doubling every decade. More important, LNG is readily available from global sources through a well-established delivery mechanism, with a resource base big enough to last for more than 200 years at current rates of demand growth.

It would be in India's interests—as it becomes as significant to global markets as the United States was before its shale bonanza—to have a diversified supply, a balanced primary energy mix and, if possible, a level of redundancy. For that reason, well-functioning, liberalized global energy market is in India's interest and worthy of attention for its energy diplomacy and foreign policy.

India has often been frustrated when dealing with energy bodies, and there is no overriding global governance for international energy markets. So, India should work with other second-tier energy consumers, especially among Asia-Pacific countries, to shape institutions or create a regional energy regime.

The more oil, coal and gas that is available through international trade, the more secure India's energy system will be. India is not an outlier in its reliance on imports, and it should accept and then prepare for the path towards integrating with a global market. Its dealings with that market will mark no small change, as will the need for more modern institutional structures to govern all energy matters. Those dealings will also require large infrastructure investments (particularly in port capacity), and a much louder voice for efficient energy markets in its foreign policy.

Energy for an urbanizing India

The world's cities hold 3.6 billion of its people, roughly half the world's population. By 2050 the number could increase to 6.3 billion—around three-quarters of the world's population. Just three countries—India, China and Nigeria—together are expected to account for 37% of the projected growth of the world's urban population between 2014 and 2050.

As the number of cities and the size of these cities grow, so will their footprint and use of energy, water and food—demand for all three could rise by 40–50% by 2030. Global demand for energy could rise by 80% by the middle of the century, with the demand coming from cities up from 66% today to 80% by 2040.

These projections have implications for the way cities are designed and managed. With so much future resource demand concentrated in urban hubs and with so much incremental demand to be met, urbanization presents a huge opportunity to become more efficient in the use of energy, water and land. Denser well-planned cities could be much more efficient than those that are unplanned and sparsely spread out. Urbanization, if planned and organized badly, risks stressing resources further. To quote Shell and the Singapore Centre for Livable Cities' New Lens on Cities:

> Well-designed and managed cities can act as powerful engines for economic development and prosperity—and help to nurture innovation and collaboration. But cities that develop poorly affect quality of life, have negative environmental impacts including higher greenhouse gas emissions, and can be the source of social and political strife.

The genuine underlying user demand is for the kilometres driven, and not for litres of petrol consumed. In Shell's Mountains scenario, centralized government and effective policy feed through into better designed (and smarter) cities, and inhabitants can live closer to their place of work or use fully integrated public transport systems. This lowers the vehicle kilometres driven by around 2,000 kilometres per person a year from low-density development in many other parts of the world (Figure 1.12).

Figure 1.12.

Smarter city planning in India can reduce transport demand

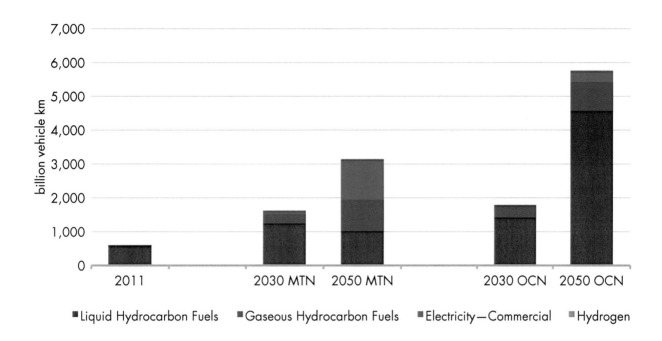

There is no single correct method or approach, but with India's being one of the main engines of global urban growth and with its well-documented resource stresses, it must make urban planning as one of its top priorities for building an efficient and resilient energy system. The recent Smart Cities initiative must focus on evaluating integrated and innovative solutions to urban planning, keeping in mind the India-specific needs and contexts.[7] Appropriate and low-carbon development to improve livelihoods in rural areas and the creation of many more centres of dynamic urban development could reduce pressures for migrating to existing urban mega-agglomerations.

Reducing energy poverty with cleaner fuels

Even with India's rising incomes, 21% of the population lives below the poverty line. Nearly two-thirds of the households use fuels such as firewood, dung-cake and charcoal for cooking. Rising wealth will usher in a shift from traditional fuels to cleaner-burning, more modern fuels such as LPG and electricity. Making these fuels available for the lowest income rungs will reduce energy poverty. The focus must be on ensuring the supply of cleaner fuels for all levels of society and lifting as many as possible out of energy poverty.

Much energy poverty is linked to energy access. India has deep divides in the quantity and quality of energy consumed across income groups and between rural and urban households. Despite six decades of independence and more than two decades of economic liberalization, energy access remains grim: nearly 20% of households consume no electricity, almost 94% of them in rural areas. For cooking fuels, nearly two-thirds of households consume no LPG, 85% of them in rural areas, with firewood as the predominant (possibly preferred) cooking fuel. A survey on multidimensional energy poverty in India, conducted by CEEW and Columbia University, found that only 5% of rural households exclusively used LPG for cooking.[8] Sustained use of traditional biomass has terrible consequences for the quality of indoor air and the health of women and children, often more exposed to the emissions.

As per capita incomes rise in rural areas, the fraction spent on energy declines (Figure 1.13). And as people move up the income ladder, they prefer more efficient fuels like LPG and electricity.

Figure 1.13.

Spending on fuel in rural areas falls as incomes rise

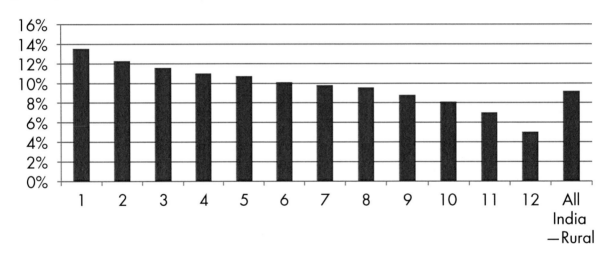

Note: Group 1 has the lowest monthly per capita expenditure, and group 12 the highest.[9]

How might fuel use in domestic cooking and water heating evolve? The Mountains scenario is an environment of strong state intervention and large pan-country initiatives. Oceans rely more on a bottom-up, disaggregated, distributed response (Figure 1.14). Both scenarios see a reduction in traditional biomass as incomes rise, so this fuel use should decline. But in Mountains clean cooking technology means that biomass is cleaner and more efficient. In Oceans stronger economic growth and higher energy prices lead to greater proliferation of distributed solar photovoltaics, with their off-grid capabilities also attractive in a world of fewer large infrastructure projects and faltering grid improvements.

Similarly, TERI's analysis indicates the need for multiple solutions targeting different user groups. TERI and CEEW's rural energy surveys reveal that even if households had LPG available, many resorted to fuel stacking for cooking solutions and continue to use biomass as well. This clearly indicates that LPG may not be the best solution for those who would like to continue using biomass for various reasons, and improved cookstoves may be as relevant a solution for a large section of people.[10]

Figure 1.14.

Traditional biomass for domestic cooking and water heating declines as incomes rise

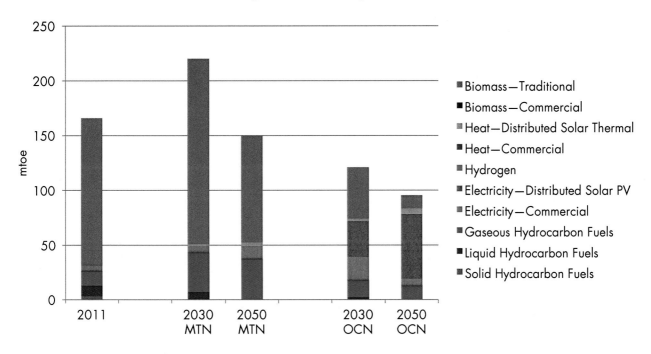

In TERI's alternative RES and UCG scenarios the final energy demand of the residential sector grows to 242 mtoe by 2050, about 26% lower than the demand in the BAU scenario. The difference is in electricity use, which by 2030 is 23% lower than in the BAU scenario. This reduction occurs due to higher penetration (100% by 2050) and use of efficient household appliances, such as air conditioners, refrigerators and LED lighting. But traditional biomass remains the most popular fuel choice, especially in rural households, although the penetration of improved cookstoves is assumed in both scenarios to reach 50% by 2050. Improved cookstoves are up to five times more efficient than traditional variants, bringing health benefits as well as energy savings. This analysis clearly indicates the role of technology progress in reducing energy demand along with appropriate changes to policy and regulatory frameworks.

The final energy demand of the commercial sector in the BAU scenario grows 19 times, from 16 mtoe in 2011 to 295 mtoe by 2050. Petroleum products and electricity are the two most popular fuel choices.

In the RES and UCG scenarios the final energy demand rises to 249 mtoe by 2050, 16% less than in BAU. A fall in electricity use is also seen, with savings of 20%. The reduction occurs due the assumption of a 5% reduction in the Environmental Performance Index every five years from 2011 for air-conditioned buildings. An increase in penetration of GRIHA (Green Rating for Integrated Habitat Assessment)-rated buildings in new-built areas is also considered (from 1% in 2011 to 13% in 2021 and 26% by 2030).

The analysis shows that, while there are different pathways of getting affordable energy to the population, they all demonstrate a shift from traditional fuels to modern fuels. How India facilitates this process will reflect policy preferences and the right pricing regime, but priorities should be reframed to cater to the different (consumption and production) energy demands of citizens of various economic strata and to direct energy subsidies more efficiently to the poor.[11]

To meet its developmental growth targets, India needs to make rapid progress on technological changes, but it must do so in a clearly planned and phased manner—supported by a well-structured policy and institutional setup, greater integration with world markets and a more holistic and integrated energy and infrastructure planning system.

Preparing for a decarbonizing world

India may also find itself in the midst of conditions not entirely of its own making. With the global population growing to 9 billion by 2050 and huge numbers rising out of poverty, developing countries, such as China and India, could more than double their final energy demand over 2000 levels and account for around two-thirds of global energy use by 2050.

This growth in energy demand comes with huge challenges, among them is the rate of emitting carbon dioxide into the atmosphere. Meeting global—particularly developing-country—energy demand is a fundamental requirement to continue building a better quality of life for citizens. But carbon emission trajectories show, under both Mountains or Oceans assumptions, that global warming will not be kept to a rise of 2 °C (Figure 1.15).

India's emission pathways are not for debate here. But policymakers must be aware that many countries will be reducing the emissions their energy systems produce. More than a few developed countries will be in a strong position to do so, with their economies now slanted more towards services and personal consumption than industry. India has an opportunity, though, to enter partnerships and agreements with these leading countries to benefit from their learning and technology for cleaner energy, without compromising its own economic growth plans.

Figure 1.15.

Emission profiles across scenarios

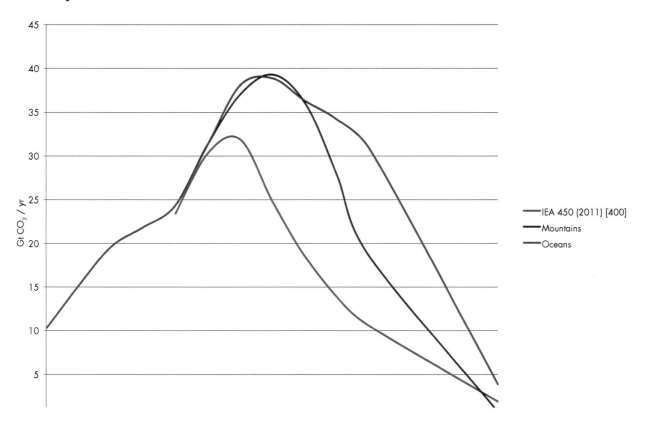

Conclusions

India's population and economic growth point to stresses on its energy system over the next 35 years, but the country should not be hoping for a silver bullet. There isn't one. Instead, both supply-side and demand-side measures must be put in place for incremental change.

Three final points. First, India's future energy demand will be large, no matter the projection. India will continue to have a significant requirement for oil, gas and coal, and so needs to adjust to the step changes it will have to make in dealing with global energy markets and in building institutional structures. Second, it has a real opportunity to put in place forward-looking policies for urbanizing efficiently and for managing the associated stresses in energy, food, water and land. Third, the debate on energy access should be focused on lifting people out of energy poverty and shifting from traditional to more efficient and cleaner burning fuels. Technologies can ease this transition. In addition, energy policy should focus on the following objectives: reducing energy poverty by providing energy to every Indian, securing stable energy supplies and limiting carbon emissions without stunting economic growth.

CHAPTER 2

Infrastructure for an Integrated Energy System

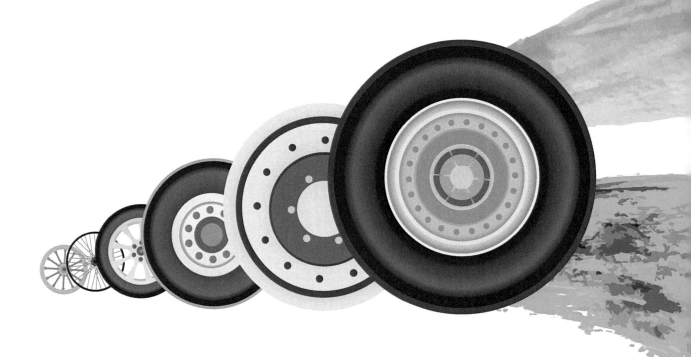

Even in a scenario with intensive efforts to shift the energy system to a low carbon trajectory, fossil fuels will play a big role. These fuels have been the backbone of the energy system, and planning for augmenting capacity has been firmly in the hands of the public sector. But the scale of infrastructure to provide for the rising energy demand and to enable a transition in the energy system will need more coordination between the public and private sectors.

A simple illustration of the shifts in planning for long-term infrastructure is visible in the (power generation) developments proposed in the 10th and 11th Five Year Plans and how they panned out—and the outlook in the 12th Plan for 2012–2017 (Table 2.1). Although the illustration is for electricity, it typifies the continuous changes across different elements of the energy system—changes that may completely revamp the landscape in the next decade.

Table 2.1.
Thermal-based generation plans

	10th FYP (2002–2007)		11th FYP (2007–2012)		12th FYP (2012–2017)
	Planned (MW)	**Achieved** (MW)	**Planned** (MW)	**Achieved** (MW)	**Planned** (MW)
Centre	12,790	6,590	24,480	12,790	14,878
State	6,675	3,553	23,301	14,030	13,922
Private	5,951	1,970	11,552	21,270	43,540

The key feature is the growing scale of the demands made on each player. At lower levels of demand and with a very strict planning (and resource allocation regime), the role of the government was immense. As indicated by the achievements over the plan periods, while the private sector had difficulty meeting its capacity quota in the 10th Plan period, it has outperformed the public sector in recent years and the expectations for it in the years ahead are also significant. Concomitantly, the role of the public sector (both central and state) is likely to decline with the increasing tilt towards renewables, although the public sector would remain the prime mover for hydro and nuclear-based generation. These would continue to be needed in the short to medium term to support the transition in power generation.

Even in the upstream energy sectors—for the production of coal and gas—more coordination is needed to ensure that private sector demands and the terms of their contracts are met. In coal, the fuel supply agreements are not fully honoured, and in gas, because of preferential allocations, the power sector ends up suffering when there is a shortfall in supply. Downstream, the transmission infrastructure has not come at the pace required. Until the 11th

FYP, growth in the transmission network (measured in circuit kilometres) was higher than the growth in generation capacity (measured in MW). But in subsequent years, generation capacity has grown at more than twice the rate of the transmission network.[12] Interestingly, renewable energy, to be a pillar in power sector growth in the years ahead, is almost entirely in private hands. And its success is likely to be severely hampered unless there is proper planning for the requisite transmission infrastructure.

Until the last decade the main drivers of energy policy were the two key objectives pursued by successive governments—energy access and energy security. Providing energy access to India's entire population has been a top priority of policymakers for a long time, making it equally or even more important than energy security. With dedicated programmes to tackle energy access and a move away from autarkic definitions of energy security in the globalized world of energy production and consumption, the focus must firmly shift to strengthening the supply backbone to support India's rapid and robust growth in energy consumption.

Building and using infrastructure has traditionally been carbon-intensive—power generation, urban amenities and transport have particularly broad carbon footprints. India has taken a voluntary (though non-binding) decision to reduce its emission intensity. In addition, the government has over the last year been aggressively positioning for 175 GW of installed capacity by 2022. Achieving this target will require resources to be mobilized at a pace never before achieved. So developing infrastructure in a manner that will be sustainable while staying aligned with India's growth, energy security and poverty alleviation objectives will be a major challenge over the next few decades.

The World Economic Forum, while ranking India 56th (of 142 in 2011) in the ease of doing business rankings, suggested that inadequate supply of infrastructure was the single biggest bottleneck for doing business in the country. The barriers range from financial constraints to cultural behaviours and to the lack of appropriate institutions for reigning in energy policies under one umbrella and, most important, to the need to transition to an energy system that looks very different from what India has in place today.

Energy consumption and the infrastructure required

As part of the effort to illustrate the scale of infrastructure required, two scenarios were evaluated using CEEW's India Energy Model (INEM), which are used to discuss the broad energy sector infrastructure requirements in the following section. While the purpose is not to be deterministic about the level of infrastructure, the intent is to provide useful directions to infrastructure planning in India's transitions.

In the reference scenario for the analysis (W&W), the total primary energy supply rises from 700 mtoe in 2011 to approximately 2300 mtoe in 2050. Coal constitutes about 66% of the total primary energy supply (1600 mtoe, or 3 billion tons of coal) in 2050. This demand for coal is nearly six times current levels. Natural gas consumption rises fivefold by 2050 (150 mtoe or 176 billion cubic metres).[13] Power consumption increases six-fold, and the overall energy consumption by households and commercial centres only doubles by 2050, thanks to the drop in traditional biomass and the move to commercial biomass and modern fuels in households. Transportation energy demand nearly quadruples to nearly 270 mtoe by 2050, and agricultural demand more than doubles to 62 mtoe.

In the INEM alternative sustainable scenarios, India's energy demand comes down by a mere 5%, suggesting that many of the efficiency gains in Indian industry and other end use applications are likely to be internalized even in the reference scenario. Coal sees a significant reduction in use because of the constraint on overall emissions and constitutes only about 30% of the total primary energy supply. Since importing natural gas at the assumed prices is still expensive, there is little uptake of imported gas in the early periods of the analysis. But it slowly rises to reach nearly 770 bcm (around 640 mtoe) by 2050, 15 times that in the base year 2011. Even with more abundant and cheaper biomass (a competing fuel), natural gas consumption grows nearly 12 times. The preferred fuel source for the power sector, it accounts for 35% of the electricity generation by 2050.

Non-fossil resources such as hydro-power, nuclear and renewables account for another third of the electricity generation. Renewables (excluding large hydro, as defined by India) account for a little less than a quarter of the overall electricity generation. There is a big shift away from fossil fuel-based generation. The transportation sector reduces energy use by more than 20% between the two scenarios, thanks to the greater efficiency associated with electricity-based traction—both in public and private vehicles.

What does this mean for the infrastructure that will have to be created to extract, transport and consume India's energy resources? The two scenarios allude to two significantly different energy systems where the contribution shifts from fossil fuels to renewable alternatives.

Coal accounts for nearly 50% of the freight movement on the rail network, and the coal to be handled in the network is expected to be nearly twice that handled in 2011. Most is carried on the Howrah–Delhi and Howrah–Mumbai routes. Both have been overused and not been running efficiently. Additional production from three major mines in Chhattisgarh, Odisha and Jharkhand will add substantially to the overall freight movements on the rail network. While the dedicated freight corridor is expected to alleviate some of the stress on these congested routes, progress has not been encouraging. More important, more than 70% of the proven coal reserves are in Jharkhand, Chattisgarh and Orissa, served by the Karanpura, Korba, Ib Valley and the Talcher fields. Increasing capacity to bring on board the coal mined from these fields also faces challenges from clearances for infrastructure projects in the forested tracts in these states. TERI's analysis indicates that the cost of underground coal mining is much higher compared to that of imported coal. Moreover, given the need to wash Indian coal, the need to set up additional washery capacities could also be avoided if the country were to rely on imported coal in supporting its requirements in the interim (transition) phase. While additional port capacities or transportation capacities could still be diverted to other uses, infrastructure and technology for underground mining and coal washing could prove to be unproductive investments in the future. The need to examine infrastructure lock-in prospects carefully is therefore required. While port capacity can be diverted to other bulk cargo, the investments need to factor in changes in requirements over time.

In the alternative "sustainable view" of the energy system (INEM), electricity production relies on coal for 40% of the contribution, and the location and variability of the new generation sources pose different challenges for electricity transmission infrastructure with resources distributed unequally over the different regions of the country. The southern region of the country could shoulder nearly a third of the renewables capacity installed by 2030, but will have only a little more than 20% of the installed fossil fuel-based capacity. The southern region is also likely to see power deficits in 2030, so interregional transmission capacity has to be augmented significantly to provide for the demands of every part of the country.

The uptake of natural gas also increases in this alternative scenario and the pipelines will handle 40% more, while coal transport looks relatively unchanged between the two scenarios. India has a gas supply network of about 15,000 km with a designed capacity of 151 billion cubic metres annually.[14]

The coverage of the pipeline network determines where the gas would get consumed and accordingly limits the maximum availability to end-user capacity. The gas pipeline network currently is highly uneven and precludes access to most of the states due to a lack of connectivity with source. A massive 79% of total gas is consumed in the western and northern parts of country, due to the lack of connectivity in the other parts. This tempers the realization of the full demand for natural gas in the country, as many potential demand centres remain devoid of gas supply due to pipeline unavailability, therefore, planning for additional gas pipelines or deciding to go for LNG requires due consideration.

Infrastructure and lock-in

Infrastructure investments today will determine whether India will be able to tread a low-carbon growth path tomorrow. The energy infrastructure capacity needed will roughly double every decade as per INEM. If so, the current installed capacity would constitute only a quarter of the capacity in 20 years (less with retirements and obsolescence). So the energy mix can change substantially over this timeframe, if incremental investments vary significantly from the installed base. That makes it important to understand the different types of infrastructure lock-ins that can arise and combat them with suitable actions.[15]

Policymakers should understand the three broad categories of lock-in that infrastructure is susceptible to.

- The first, technological lock-in, is related to the life of the equipment and the rate of new investment. Companies that benefit from existing technologies may actively try to prevent the emergence of new technologies in the market, either through market action or through influencing the policy process. The sustained use of coal and gas fired power units or the propagation of the urban form as envisioned in the 1960s in the United States are examples. For India a grid that is not prepared to handle variability in generation perpetuates the need to rely exclusively on baseload type steady generation sources.
- Second is preference lock-in. Large infrastructure investments tend to shape individual and societal preferences, where people link certain levels of service with a specific way of life and are unable to consider alternatives that might deliver the same at a lower environmental footprint. India's mega-cities could be a result of an innate preference for the existing centres and the agglomerations around them.
- A third kind of lock-in emanates from the political economy, which often facilitates technological lock-in and preference lock-in. Even where solutions are available, there may be impediments to adoption because of the differential ability of the affected parties to influence the policy process.[16] Subsidies for the consumption of fossil fuels preclude even the wealthy from considering the impacts of their consumption decisions. And despite the large electricity deficit today, most natural gas-based generation stations are severely underused and do not find takers for the electricity they generate because of the cost.

The solutions to infrastructure lock-in are not obvious. They require breaking established preferences and choices. Only a healthy dialogue among the various stakeholders can identify the right investment strategy to avoid technology lock-in and consider how today's policy choices will affect the ability to alter them later.

Regulating India's future energy infrastructure

The command and control mode of governance that relied on state ownership of infrastructure services is gradually moving towards "government as regulator." Public–private partnerships and private sector participation require that government achieves its priorities through independent regulation and through laws and contracts, a shift that remains inadequately understood.

In the energy sector, the Electricity Act (2003), Coal Mines Nationalisation Act (1973) and the Petroleum and Natural Gas Regulatory Board Act (2006) are pivotal in the regulation of the sectors they are responsible for. The Electricity Act created central and state regulatory commissions that are unable to direct the developments in the sector without interference from the executive branch. For coal the coal controller does not have the independence of a regulator, and the ministry directly oversees the monopoly producer (also a public sector undertaking). For gas, more than 15 years of private sector involvement (through the New Exploration Licensing Policy) has resulted in a total investment of only $21 billion.

How can key issues that determine the effectiveness of regulatory institutions be resolved to make regulation meaningful and efficient? By paying heed to the federal principle (intrinsic to the functioning of any governance initiative in India), by separating powers (separate judicial power for an independent entity at the very least) and by ensuring democratic accountability (to the people and through the legislature, as opposed to the executive).

Laws

Future infrastructure demands a fine balance between economic development and environmental protection, along with maintaining social safeguards. In India "green clearances" are critical instruments to balance this trade-off. A delay in commissioning a single project usually has a serial impact on all the associated activities and investments. Consider the recently commissioned Kochi RLNG terminal, which is riddled with financial distress due to a lack of gas transmission infrastructure (specifically the proposed Kochi–Koottanad–Bangalore–Mangalore pipeline) that would actually deliver gas to large consumer markets down the supply chain. Similarly, India's ambition to ramp up oil production has also faced delayed clearances. Indeed, investments worth $12.4 billion in upstream hydrocarbon exploration were stalled for want of clearances from multiple ministries and state governments.

Across a range of infrastructure projects, most of the applications filed between 2003 and 2014, for new projects and capacity expansions, were granted environmental clearance. A CEEW analysis of more than 11,100 projects (for their environmental clearance status) reveals some interesting findings (Figure 2.1). Many projects across multiple sectors, especially in the industrial category (90%), are still awaiting forest clearances. Nearly 52% of the project applications filed in 2010 were still awaiting forest clearances in August 2014.[17]

Even if acquiring land has no social reservations, the process still takes up to 50 months (without accounting for extensions).[18] Similarly, adhering to prescribed environment clearance process could take up to 940 days (including the time for environmental impact assessments and other scoping studies). The uncertainty of each individual clearance makes things even more daunting, since failing to obtain one clearance makes all other efforts wasteful.

Regulatory delays, often a result of red-tape, disturb investment sentiment, evident in the withdrawal of major oil and gas exploration companies from India. India ranks poorly (142nd) in the World Bank's Doing Business reports. The current administration is trying to restore investor confidence through its ambitious "Make in India" campaign. The recent introduction of an online portal for project clearances is a step in the right direction.

Figure 2.1.

Approved projects awaiting forest clearances across various sectors, 2003–2014

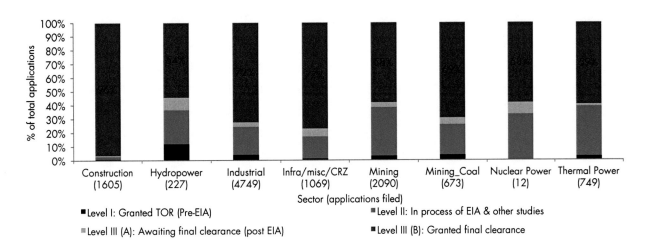

Source: CEEW analysis based on primary data from the Ministry of Environmental, Forest and Climate Change 2014.

Regulating energy infrastructure through contracts

Concession agreements and contracts signed by public authorities and private entities govern several infrastructure projects and services. The agreements also set tariffs and performance standards that typically are subjects of independent regulation. A well-defined contract can lead to greater predictability and enforceability for provision of infrastructure services—a flawed contract, to the opposite. Prior consultations with the regulator and stakeholders can help create an appropriate contractual framework and eliminate unintended outcomes. A process for standardizing concessions and contracts should be part of the overarching approach to regulation.

Contracts are the basis for project viability and they control the allocation of risks. Companies responsible for carrying out projects have a series of contracts to unite various parties in a vertical chain from input supplier to output purchaser. Even though contracts play an enormous role in carrying out projects, their use and application for aligning projects with the goals of a sustainable and efficient energy system have been limited—for a variety of reasons.

There is almost no voluntary adoption of good environmental management practices among contractors. So the obligations under a contract become the main tool to ensure that good practices are implemented on the ground. Some initiatives, such as a specific environmental management plan, are taken only because of contract requirements.

Among the most important reasons for India's rapid strides in private participation in infrastructure is drawing up model concession agreements, which define the risk–reward framework for stakeholders in precise terms. Standardized documents from international best practices lend transparency to, and expedite, the project award and implementation process. But in recent times the Central Electricity Regulatory Commission granted relief to two imported coal-based power projects by agreeing to reopen signed long-term power purchase agreements. Many road projects, including the recently awarded Kishangarh–Udaipur–Ahmedabad project (to a big infrastructure company), are also seeking relief after contract award through renegotiations. Renegotiations compromise the sanctity of contracts. Everything would be negotiable even after contract award, forsaking the benefits of a competitive auction. In evaluating these requests, regulators should distinguish between renegotiation demands based on incomplete contracts and those based on opportunistic behaviour—and consider only the former.

Putting together the pieces

In the INEM W&W scenario, the three most important elements of the energy mix in India are coal, gas and renewables, contributing to more than 80% of the primary energy mix. Even in a scenario where India attempts to aggressively limit CO_2 emissions (LCIG), the contribution by these sources remains high, especially with the growth in demand for transportation.

Coal

Challenges for domestic coal production arise from many considerations ranging from availability of credible data on extractable reserves to monopoly in the production of coal to the difficulty in land acquisition and appropriate compensation and rehabilitation of affected populations.

Even if the required coal is available (domestic and imported), infrastructure has to support its movement. Coal handling capacity at major and minor ports, along with availability of supporting infrastructure from Indian railways, becomes crucial. Mines in Ib Valley, Karanpura, Mand-Raigarh and elsewhere in India, if exploited well, can ease the domestic supply situation to the tune of 100 MT to 200 MT. Given that coal needs to be part of the solution, planning the infrastructure for extraction, processing, transportation and use is important.

There needs to be an independent coal regulator (currently a role performed by the Government of India). An independent, empowered and accountable regulator would help in addressing a fundamental conflict of interest in the sector, where the Government of India is both a majority shareholder in Coal India Limited (CIL) and the de-facto regulator as a custodian of the national resource. Central Mine Planning & Design Institute Limited, the primary coal exploration agency, must be made autonomous from producing entities such as CIL.

Also important is building the human resource capacity at agencies such as Central Mine Planning & Design Institute Limited and improving the functioning of the coal sector, which does not attract good talent. Assessing the resource base better and evaluating the technological feasibility of underground mining would serve the industry well in the long run.

At least in the short term, India's coal imports will increase significantly, and India may soon be one of the world's largest coal importers. It must thus focus on ensuring that the requisite port capacity is available to accommodate the large increase in demand in the next decade.

Natural gas

India's natural gas transition, though initially dependent on domestic resources, is likely to thrive only in a scenario where imports significantly overshadow domestic production. With imports primarily from the Persian Gulf, the United States and Australia, the main concern for policymakers must be to encourage the creation of new regasified liquefied natural gas facilities to augment imports.

It is important not only to have enough natural gas reserves and production, but also to ensure a steady flow of gas from producing wells and regasified liquefied natural gas terminals to the end-users. India fails when it comes to cross-country gas pipelines, which at present cover only about 15,000 km. Regional imbalances are largely due to gas markets remaining in places where gas sources are found, deterring further expansion (Table 2.2).

The Kochi–Koottanad pipeline cited earlier is a case in point. The low capacity use of existing pipelines, currently averaging 38%, has discouraged investments in gas supply infrastructure and rendered it unprofitable. A Gas Authority of India Limited plan to invest $3 billion in new pipelines has been deferred. The reasons: the unviability of regasified liquefied natural gas and the declining production from domestic fields in recent years.[19]

This has resulted in financial losses not only to the regasified liquefied natural gas facility operator but also to investors in power plants and fertilizer units anticipating gas from the source locations. If the demand for gas in the country is to be met in the years ahead, increasing LNG import capacity (regasification facilities) is essential. Despite long pending plan of building transnational pipelines, LNG will likely be the main source of imported natural gas for the short and medium term. If investors are required to invest in building more of these terminals, the vicious cycle associated with natural gas needs to be broken with the right interventions.

Table 2.2.

Regional distribution of the gas pipeline network in India

Region	% of consumption	States with pipeline infrastructure	States lacking connectivity with pipeline infrastructure
Western	53%	Gujarat, Maharashtra, Goa	–
Northern	26%	Delhi, UP, Haryana, Rajasthan, Punjab	Jammu & Kashmir, Himachal Pradesh, Uttarakhand
Central	3%	Madhya Pradesh	Chhattisgarh
Southern	14%	Andhra Pradesh, Karnataka	Tamil Nadu, Kerala
Eastern	Nil		Bihar, West Bengal, Jharkhand, Odisha
North Eastern	4%	Assam, Tripura	Meghalaya, Sikkim, Arunachal Pradesh, Mizoram, Manipur, Nagaland

Note: The states of Northeast region mainly have regional gas networks, sourcing gas from domestic production fields. They are not connected to the national gas network.

Source: Adapted from PNGRB 2013.

Renewables

While coal and natural gas form the backbone for power and industrial use, India has ambitious targets for renewable energy growth. As part of its intended contribution towards mitigating climate change, India aims to install an additional 75 GW of wind power capacity and 100 GW of solar power capacity by 2022. This is more than six times the current installed capacities of about 22 GW and 5 GW, for wind and solar respectively.

In India there has been policy support over the last decade for renewables, but some specifics have not been stable. Accelerated depreciation for wind power was withdrawn and feed-in-tariffs have fallen short of investor expectations. The support needs to be more comprehensive rather than piecemeal.

Most acquisition of private land for wind and solar projects occurs through mediations directly with the land owners, a major barrier for renewable projects across India—indeed, all power projects. And the lead time to acquire land can range from 6 to 12 months to more than a year. There should thus be single-window clearance for all approvals. Land acquisition should become easier through faster forest approvals and use classifications.

To further support the growth of renewables, the expansion of transmission infrastructure is required. Grid enhancement and management are needed to integrate variable renewable power. Large-scale wind and solar integration, if not adequately planned, could cause congestion and imbalances in the transmission and distribution networks. India needs to therefore develop a robust interstate transmission network to evacuate power. The Ministry of New and Renewable Energy and Power Grid Corporation of India Limited (PGCIL) have planned six dedicated green energy corridors to evacuate renewable energy and feed it to other regions. These corridors can also address intermittency, variability and the grid integration of large-scale renewable generation.

The integration of variable renewable power also requires better scheduling and forecasting of the various sources that contribute to it. India should set up state-of-the-art centralized forecasting centres, which will have to be integrated with supervisory control and data acquisition systems. There has been a draft proposal to set up an ancillary services market in India—but progress has been at a minimum. Without a robust market that delivers peaking power, frequency support and voltage support, it is unlikely that the ambitious plans to take on climate change will pay off.

Domestic manufacturing and research and development

For thermal generation, India has at last managed to establish a critical mass of manufacturing of super-critical boilers and related components to power the next wave of electricity generation facilities. Nearly 35,000 MW of annual production capacity of supercritical boilers have been established through licensing and joint-ventures with overseas entities. India is also independently conducting R&D on advance ultra-supercritical boilers so that it does not become overly reliant on overseas entities for licensing.

For solar technology there is little investment in customizing solutions for the Indian environment. Government and industry need to develop substantial R&D capabilities to suit the needs of the wind and solar power sectors. This can also help keep costs under control, especially for the balance of systems (components other than the panel itself), which constitute nearly 50% of the overall cost of a project.

In addition to the lack of vertically integrated production lines (for core solar PV components), Indian firms have also faced competitive industries that supply balance of systems, like power inverters and transformers to solar power plants. Indian firms are highly reliant on turnkey manufacturing, signalling that the industry has lagged behind in innovation and process engineering.[19] Despite being early movers, Indian manufacturers were unable to capitalize on their position. They failed to adapt to the global changes and challenges in the sector, partly due to the inherent deficiencies of the manufacturing ecosystem in India and the failure of institutions and policies meant to cater to the solar PV industry. Various firms cried foul in the face of cheap imports but did not coordinate well with research institutions and invest in R&D.

For wind India has significant domestic capacity to manufacture towers and blades, which constitute up to 40% of the cost of all the components in a wind energy project. But even in this process the steel and fibre composites required for towers and blades are in many cases imported. And the remaining 60% of the components need to be imported from European bases, pushing up costs for the wind industry.

There is an urgent need to envision the future state of solar and wind technology (in the next few decades) as they come online in a big way in the generation mix. Both the private and public sectors must promote R&D that has India at the driver's seat—controlling developments and not being a mere importer of components.

Soft infrastructure

Building human capabilities

Oil and gas have benefitted from international practices and the presence of MNCs in India, while other energy sectors have remained more insulated so far in terms of global integration. Even in the power sector, which has established players in its ranks, capacity building has not been on track. Refresher training for upgrading knowledge and skills of those employed in the power sector is only 3% of what is required, and the training capacity for managerial staff only 4%.[20] A joint study by CEEW and NRDC finds that nearly one million jobs have to be created in the solar space alone (by 2022) to cater to the renewable energy targets that have been set.

India needs to focus on the requirements specific to the various components of the chain that delivers the desired workforce for the energy industry. Capacity building for the energy sector could focus on four key areas that would result in appropriate skilling of the workforce.

- **Information and staffing**. A comprehensive database should be created for projects that require staffing, for trainers and recruiting agencies and for potential employees.
- **Creating certification and degree programmes**. Apprentice schemes and vocational training courses, the strength of the traditionally industrialized states, need to be replicated to ensure that the energy sector can reap the same benefits as other industries.
- **Content creation**. Perhaps the most important element of building human capabilities is the content that will be delivered through training programmes, hence the need for content developers, for a library of resources and for meeting the required standards.
- **Training infrastructure**. In the power sector much of the trained staff has been associated with the public training schools. It will be important to create and bring to scale private infrastructure to meet the requisite number of trainers, labs and workshops. Certifying these resources and creating the right accreditation criteria are important elements to focus on.

Financing

Almost half of the total infrastructure investment in the current Five Year Plan (2012–2017) is expected to be financed by private sources. But their contribution has been limited given the regulatory restrictions on minimum credit ratings of their investments.

So the main issue in financing infrastructure is channelling long-term savings into infrastructure investment through low credit risk securities. This requires financial intermediaries with adequate due diligence, monitoring and structuring skills for infrastructure projects. The government has taken steps through the market and banking regulators to provide regulatory frameworks for specialized infrastructure financing intermediaries. It has also set up the India Infrastructure Finance Company Limited to provide long-term financing and credit enhancement for bond issues by PPP projects. And to enhance the supply of long-term financing, the government enables public sector infrastructure development companies to issue budgeted amounts of long-term tax-free infrastructure bonds to institutional and retail investors.

However, the inherent risks, such as the insolvency of electricity utilities or the risk associated with the inability of Coal India to supply coal, make it difficult for these projects to qualify for these novel financial instruments. This risk is more pronounced for creating renewable energy assets since there is also a component of technology risk, as they have yet to demonstrate their long-term performance (over 20 to 25 years).

Banks can at best be expected to provide short-term financing during the construction period. It has also been shown that tax-exempt bonds issued by government end up being more expensive than other instruments available in the market, such as partial risk guarantees. To the extent that pension and insurance funds do not have their own due diligence capabilities for infrastructure projects, infrastructure finance companies and infrastructure debt funds can provide such services. The bulk of the funding then will have to come from the emerging bond markets. If the infrastructure projects do not meet the minimum ratings requirements, the Infrastructure Investment and Financing Company Limited can provide credit enhancement to the bonds issued by such projects.

Conclusions

India's transition to a lower carbon economy over the next 30 to 35 years will happen at a slow pace. But it can make the transition by developing, commercializing and integrating known but currently underdeveloped solutions. In the decade ahead, India's energy policy should focus on preparedness. It has to develop options and explore trade-offs, while also testing our technical, operating, business and regulatory models at a sufficient scale to give stakeholders the confidence they need to commit to full-scale implementation.

CHAPTER 3

Technology for a Productive Energy System

Transitioning to a different energy and technology mix is inevitable for India's sustainable energy future. This inevitability arises not only from the implications that the large and growing energy requirements could have on the country's land, water and air resources, but also in terms of the economic implications and infrastructural requirements for handling, transporting and distributing energy across the country. Choosing appropriate technologies at various levels and across the entire energy value chain is thus a key element in India's energy future. Since energy-related infrastructure is generally associated with large gestation periods as also fairly long economic lifetimes (often around 25–30 years at least) of energy infrastructure, machinery and equipment, the need to develop, adopt, adapt and up-scale appropriate technology choices—ranging from clean fuel production technologies to resource conversion technologies and to end-use technologies—in a timely and well-planned manner cannot be undermined.

As indicated by the range of energy estimates for 2030 and 2050, the growth in India's final energy requirements implies that no single energy fuel or technology can be the solution. A variety of fuels and technologies will be needed across all sectors in the quest to provide adequate, reliable and affordable energy to all. Technology choices are needed across all links in the energy value chain—in resource assessment and production, in conversion, in transmission and distribution and on the demand side. Any technology to be deployed in these areas should be ranked on three important criteria: its availability, its cost-effectiveness and its environmental sustainability.

India faces significant challenges in deploying available, cost-effective and environmentally sustainable technology in each of the four links in the value chain. India's technology roadmap must focus on these criteria and plan holistically for appropriate solutions in the short, medium and long term.

The production of fossil fuels in India has been stagnating over time. The annual growth in coal production dropped from 6% a year at the end of the 10th plan period (2002–2007) to about 1.5% a year in 2013–2014, while consumption increased at an average of 7% a year. Crude oil production has also stagnated at around 40 million tonnes a year while consumption increased around 5% a year. Given that all scenarios indicate a continued need for coal, oil and gas in India's energy sector, resource assessment, exploration and production technologies must be scrutinized to examine the resource base and enable efficient extraction and distribution.

India's extractable coal reserves are not only limited but also high in ash content (ranging from 35% to 45% ash), and underground mining is expected to raise the coal price to economically unattractive levels. So plans for the production, processing (washing) and supply of indigenous coal should be carefully examined—keeping in mind the technology and infrastructure requirements of increased indigenous (high ash) coal production.

The oil and gas sector needs the same approach. The success rate of various New Exploration Licensing Policy rounds and high import dependence have time and again emphasized the inadequacy of technology deployment.

While renewables and other alternatives have to step in to meet India's energy requirements in the long run, improvements in efficiencies of extraction, resource assessments and supply of fossil fuels are required to support India's energy transition till 2050.

Currently most of India's coal power plants are subcritical technology. The average efficiency of coal thermal plants in India is 34% and that of gas thermal plants is 46%, far lower than the OECD average of 38% for coal and 50% for gas. Planning for continued and rapid improvement in efficiencies of coal and gas-based power plants is equally important in the immediate short term given the continued reliance on coal and gas-based plants indicated in all scenarios for the next couple of decades. While India has already moved to super critical technology for all new coal-based power plants, a quick transition to more advanced clean coal technologies and more efficient gas-based generation technologies is desirable for enabling a shift to lower emission intensities in India's power generation sector. Retiring inefficient and obsolete sub-critical coal-based capacities could not only enhance overall efficiency of the sector, but also work well by making available space for setting up new capacities at these locations.

Moreover, India has yet to tap its renewable resources to any significant level. It has only 23.4 GW of wind power, mostly at a hub-height of 30–50 metres, far below the assessed potential of 500 GW at 80 metres and even more at 120 metres. Solar installations currently stand at only 5 GW and modern renewable resources have so far remained largely untapped. Given, the recent voluntary commitment of 40% non-fossil capacity by 2030 in India's INDC, the need to focus attention on accelerating renewable technology development and uptake has increased manifold. RD&D efforts need to scale up rapidly to bring in developments especially with regard to batteries and storage capabilities, efficiencies of wind and solar applications need to improve further, and applications need to increasingly align to requirements of different user groups. At the same time, hydro which has historically contributed to a significant share of India's power generation capacity in the past, needs to once again be reassessed and its uptake enhanced to enable a diversified and clean energy mix in the coming decades.

Nearly 80% of households depend on traditional biomass for heating and cooking, which amounts to almost a third of the total primary energy supply. Moreover nearly 300 million people in India lack access to electricity. While providing modern energy such as LPG to meet this latent demand as well as catering to the needs of a growing population whose unmet demands are yet to be fulfilled, we need to be mindful of the fact that a significant share of people would continue to rely on biomass-based cooking even in 2050. Therefore, technological development needs to simultaneously address the needs of this section of society as well, and look at developing improved cookstoves and other affordable technological solutions suited to the needs of all socio-economic sections of our society. Technological development must simultaneously proceed in the category of white-goods and other electrical appliances used primarily in the residential and commercial sectors in order to contain the rapid increase in final energy requirements with economic development and improving lifestyles. Here again, the focus should be on improving and adapting technologies to achieve best efficiencies in India specific conditions—be they climate specific or socio-economic conditions. For example, some technologies may not function as efficiently in India as in other countries of the developed world as they may be designed to function in colder climates. Similarly, while some advanced technologies may be easily absorbed in developed countries, there may be socio-economic or cultural factors at play that inhibit their uptake in India. Therefore, understanding the needs of the market and designing/adapting technologies to suit our conditions is extremely important for long-term sustainability.

The transport sector in India is increasingly moving towards privatized transport with personalized vehicle population growing at an alarming rate. By 2050 the demand of the Indian transport sector alone may equal that of India's total energy demand in 2011. Over time we observe a shift from rail to more energy intensive road-based transport in all the reference scenarios, however the alternative scenarios indicate considerable scope for energy efficiency as an important element in transport sector which will be important in bringing down energy demand. Hybrid and electric vehicles as well as advanced alternative fuels like third generation biofuels and integrated hydro-pyrolysis and hydro conversion could make significant impacts. Improved energy efficiency across the end-use sectors can bring savings of 25–35%.

Key technologies required for India's energy system

Coal reserves and production

Considering the continued high reliance on coal in the energy system at least till the medium term, planning for optimal availability of coal resources, assessing the techno economics of domestic reserves vis-à-vis imports of coal, choosing between washing of domestic coal vis-à-vis using imported coal and evaluating cleaner options for coal-based generation assume high importance for ensuring sustainability. While upgrading indicated reserves into proven reserves is a continuous process, a critical assessment of the optimality of enhanced extraction and processing of domestic coal vis-à-vis imported coal based on coal reserves and quality is required. Given the high ash of Indian coals and therefore the need to beneficiate this prior to supply and use, may lead to large additional washing capacities being set up, which over time become redundant as the country ramps up its clean fuel generation. Of 306 billion tonnes of coal resources, only 42% is proven. An early up-grade and quality assessment of coal resources can help allocations across sectors. Proven reserve figures do not always sufficiently cover the techno-economic feasibility of extracting them, so deploying appropriate methodology to correctly assess the economically minable coal is critical.

Around 70% of domestic coal is used for electricity generation. But since only 10% is of prime or medium coking quality, sectors like iron and steel need to rely on imports for their needs. Only 21% of the proven resource is extractable using present mining technology. It is estimated that of the total extractable coal, about 15 billion tonnes are power grade, which would last only 30 years at current consumption rates. And with many greenfield power plants under construction, coal supplies will become even more critical, so assessing reserves and quality is of prime importance.

Coal exploration technologies have become more precise, and imaging techniques provide higher resolution. For example, coal seams under a high-density area can now be detected with higher resolution imaging. For quality assessments a new technology allows direct estimation of ash and sulphur content in coal during geophysical logging. So establishing a coal reserve assessment system, based on drilling and other exploration technologies, will directly contribute to coal resource development and production plans.

Ninety percent of domestic coal production currently comes from opencast mining, and the remaining 10% from underground mining. Opencast mining has some distinctive advantage over underground mining such as a higher percentage of extraction, a higher rate of production, extraction of thick seams under shallow cover, shorter gestation periods, lower production costs and better economies of scale. But limits on opencast mining technologies relate to the depth of the mine, the stripping ratio and the large areas required. If India is to plan for higher levels of production, it needs to not only enhance output from its opencast technologies, but also ensure that unused reserves are explored at greater depths with better underground technologies. Therefore, increasing production would require advanced but cost-effective underground mining technologies (Figure 3.1).

Opencast mining deploys heavy earth moving machinery, such as dragline, surface miners and in-pit crushing and conveying machines. In future, there would be a growing requirement for deeper opencast mining, super pits and mega mines, mining seams below existing opencast mines and high wall mining. Enhancing opencast production beyond current levels could push up mining costs because of the greater distances and poorer stripping ratios.

Underground mining includes conventional board and pillar, semi-mechanized board and pillar with side discharge loaders/load hail dumpers/universal drilling machines, mechanized board and pillar with continuous miners and mechanized powered support long wall. Short wall mining is also practised, while mechanized long wall mining has meet with little success because of geological conditions.

Since coal produced in India is high in ash content and has low heat value, it is essential to add washery capacity to improve the quality. Today's washery capacity stands at close to 145 Mt for coking coal and 112 Mt for non-coking coal. Capacity additions have been limited due to environment impacts and the difficulties in acquiring land. Indeed, during the 11th Five Year Plan (2007–2012), no washery capacity was added. And recent complaints about the quality of washed coal highlight the need to improve the technology.

TERI analysis indicates that while opencast coal production costs are more or less at par with that of imported coal, costs of underground mining are definitely much higher and unlikely to be able to compete with imported coals. Moreover, with washing of underground coal, the economics would inevitably become even more unattractive. Therefore, a re-look at the levels of coal exploitation and an in-depth assessment of plans for additional infrastructure of coal extraction and preparation is desirable.

Figure 3.1.
Technology interventions for coal extraction in three scenarios, 2011–2050

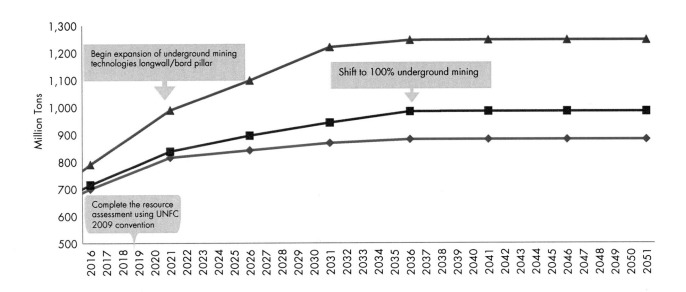

Indian coal has an average gross calorific value of about 4,500 Kcal/kg, well below the 6,500 Kcal/kg in Australia and Indonesia. As a result, Indian power plants using domestic coal consume about 0.70 kg of coal to generate 1 kWh, well above the 0.45 kg in the United States.

The high ash content of Indian coal reduces oxidation rates and thus peak furnace temperatures. It also reduces boiler loading capacities since ash takes up space and increases handling costs. This makes integrated gasification combined cycle technologies unsuitable. Therefore, India should instead look to super critical and ultra-super critical circulating fluidized bed combustion and pressurized fluidized bed combustion technologies, which can be adapted to use low grade coal with high ash. Further, research on underground coal gasification is another option that should continue to be evaluated given the country's need for relying on fossil resources for some time.

Oil and gas reserves and production

For oil and gas India must get a better understanding of the true size of its potential. Of India's sedimentary basins, 70% are poorly or not completely explored. For imaging and interpretation, India should exploit advanced technologies such as:

- Seismic acquisition,
- Seismic processing and interpretation,
- Micro-seismic remote sensing,
- Non-seismic remote sensing,
- Time-lapse methods,
- Wireline telemetry and tools,
- Rock property and rock correlation tools,
- Basin and play insight tools.

Similarly, for carbon capture and storage (CCS), more thorough assessments have to be carried out on the potential of suitable subsurface facilities. Due to the lack of data on CO_2 storage potential, most estimates are theoretical and lack adequate geological information. Moreover, adoption of CCS in India and its ability to attract attention of the Indian government depend on the successful implementation and scale-up of CCS technology in developed countries, bringing down the costs significantly. Here again, at least on the R&D front, CCS technology should continue to be studied in the Indian context, given that the country would continue to be reliant on coal and gas in the coming decades to a significant extent, even as it increases the deployment of renewables and cleaner technologies.

The 4D (as opposed to 3D) seismic technology is one which should be adopted for optimizing production. Transmitting and receiving artificially induced seismic waves for recording and analysis has long been used to assess an area's prospects by getting an impression of the rock formations below ground. For 3D seismic, a relatively recent development, results can be depicted in 3D, giving geologists far greater insight into the subsurface conditions and potential oil and gas reservoirs. For 4D the added dimension is time, with advanced processing depicting how the subsurface has been changing, thus providing a much better idea of how to optimize production.

Other technologies available to help India increase its reserves and production fall under the banner of improved oil recovery or enhanced oil recovery. They increase the volume of oil that can be produced from existing fields at an accelerated rate. Most use one of five main techniques—thermal recovery, gas injection, chemical injection, waterflood and hydraulic fracturing.

Thermal recovery uses cyclic steam stimulation or steam flooding. Injecting steam into the reservoir heats the oil while it is still in the ground to reduce its viscosity and enable it to flow more easily to a producing well. Cyclic steam stimulation injects steam into the well for some weeks or months, leaving the formation to absorb the heat before the well is opened again to pump out the oil. Steam flooding, by contrast, uses two separate wells. One is drilled horizontally above another, with the uppermost a steam injection well. As the steam is injected, the oil becomes more viscous, and gravity-assisted drainage leads it to the lower well, which sends the oil to surface.

With gas injection, gases, usually CO_2, nitrogen or lighter hydrocarbons under high pressure, are injected into the well. The gases mix with the oil, reducing its density, again making it less viscous and easier to pump.

Chemical injection reduces the capillary forces that trap residual oil in a reservoir. Because oil clings to the reservoir's rock surfaces, chemicals are injected to reduce the friction and enable more oil to be "washed" from the rocks and produced in the well.

Waterflood injects water into the reservoir to increase pressure and drive the oil towards the producing wells. Waterflood is sometimes not considered an improved oil recovery technology but simply a secondary recovery method.

Hydraulic fracturing differs in that it does not affect the viscosity of the oil but instead increases the permeability of the rock by pumping in high pressure water to crack the rock, usually with a mix of sand and guar gum to keep the fractures open. Combining this technology with horizontal drilling enables the well to reach more hydrocarbon-bearing rock. Shale gas extraction typically requires 2.8–4.8 million gallons of water per fractured well. In India the areas that could have shale gas reserves—such as Cambay, Gondwana, Krishna-Godavari and the Indo-Gangetic plains—are also vulnerable to water stress.

Conversion technologies

About 71% of India's current power capacity mix is thermal (coal, oil and gas). In June 2008, under the National Mission on Enhanced Energy Efficiency, the government launched the Perform Achieve Trade (PAT) scheme. PAT is a market-based mechanism to enhance the cost-effectiveness of improvements in energy efficiency in large energy-intensive industries and facilities by certifying the energy savings that can be traded. For the thermal power plant sector in this scheme, 144 high energy-consuming units have been notified as designated consumers by the Bureau of Energy Efficiency. Their total reported energy consumption is about 104 mtoe. By the end of the first PAT cycle (2015), energy savings of 3.211 mtoe/year are expected, around 48% of total national energy saving targets under the PAT.

Subcritical coal power technology is currently the backbone of the thermal power sector, accounting for more than 50% of the generation capacity. The power sector has attempted to diversify its coal technology, giving a push to shift to supercritical technology, which offers higher conversion efficiency. During the 12th Plan, 38% of the coal-based capacity to be installed was to be based on super critical technology, with all new power plants to be based on supercritical technology. The government should also envisage advanced ultra-super critical and integrated gasification combined cycle technology in the immediate short term in a bid to prevent lock-ins to sub-optimal technologies. This implies that barriers to the adoption and upscale of these advanced coal-based generation options be removed quickly.

The average gross efficiency of Indian coal- and lignite-fired power plants in 2009 was 33% based on a higher heating value and 34% based on a lower heating value.[21] According to a recent study by the Centre of Science and Environment under its Green Rating Project, Indian coal-based thermal power plants are some of the least efficient in the world. The technical challenges of the electricity sector include low efficiencies of thermal power plants and the continuing reliance on coal plants.

About 9% of India's generation capacity is based on gas. Current gas power technologies include both open cycle and combined cycle power plants, but most plants are combined cycle. The average net efficiency for gas fired power plants in India in 2009 was 42%, comparable to the world average in 2007. Peaking power is ideally provided by hydro reservoirs. But other suitable generation options, such as gas-based generation, could provide fast responses during peak hours. The 12th Plan envisaged developing 1,086 MW of gas-based capacity and at least 2,000 MW gas-based peaking power plants. According to TERI analysis, advanced gas (H-frame) turbines can be adopted by 2016, and these would help in saving energy and managing peak demand. Such turbines have efficiencies of more than 60% in the combined cycle operation.

About 5,780 MW of generating capacity is based on nuclear energy. India has 19 pressurized heavy water reactors and two boiling water reactors in operation. A three-stage strategy based on a closed nuclear fuel cycle has been developed. The first stage is fuelled by natural uranium to produce plutonium. The plutonium is used in the second stage in fast breeder reactors to convert thorium and uranium into fissile material. In the third stage the fissile uranium and plutonium are combined with thorium in advanced heavy water reactors, which would get about two-thirds of their fuel input from thorium. India has mastered the first stage. By 2052 India could have around 275 GW of nuclear power, with fast breeder reactors contributing around 262.5 GW.

India's nuclear reactors have been operating at 45% to 55% of capacity, due to a shortage of uranium. India does, however, have large thorium reserves. But using these resources for nuclear power generation requires a more complex chain of nuclear technologies than required for the use of uranium as fuel. India could exploit its vast thorium resources and pursue a strategy based on a closed nuclear fuel cycle.

Renewable energy generation capacity has grown from 3.5 GW in 2002 to 31.5 GW in 2015, with wind and solar behind this growth. In the Union Budget for 2015–2016, the Government announced that it would strive to achieve 175 GW of renewable energy by 2022 (100 GW for solar, 60 for wind, 5 for small hydro and 10 for biomass).

Of the two solar technologies—solar photovoltaic (PV) and concentrated solar power (CSP)—PV remains the primary technology to harness solar power in India. Most manufacturing capacity is based on crystalline silicon manufactured as wafers, and this expected to dominate the market. But thin film technology has an advantage—particularly in extreme temperature areas and where land is easily available—and is trickling in to the Indian market. Solar PV and rooftop solar can provide both centralized and decentralized energy solutions. CSP plants can be equipped with heat storage system to generate electricity, but there are no large scale CSP projects in India. Thermal storage can significantly increase CSP's dispatchability and facilitate grid integration and competitiveness. It can thus be used as base load.

Concentrated solar thermal technologies could provide low level heating requirement in large industries. They generate electricity by using mirrors to concentrate a large area of sunlight onto a small area to generate heat, which is then used to drive a turbine, usually steam. So water is an important input, and depending on the technology water could be working fluid and a heat transfer fluid, or a cleaning fluid for the solar collectors. In India, where areas with high potential for solar power are also highly water stressed, access to water could constrain the technology's deployment. Dry cooling would reduce water requirement but would require higher efficiency and low costs.

All of India's wind energy capacity (23.4 GW) is based on on-shore technology.[22] The average capacity of wind turbines in India is 20% today and is projected to increase to 30% by 2030. The capacity factor of an off-shore wind turbine is usually higher than that of an on-shore turbine, ranging from 40% to 50%. India has a vast coastline, so the use of off-shore wind turbines for electricity generation is a promising option.

The technologies used for biomass power are combustion and cogeneration. The cycle is the conventional Rankine cycle, with biomass burned in a high pressure boiler to generate steam and operate a turbine. The net power cycle efficiencies are about 23–25%. The exhaust of the steam turbine can either be fully condensed or used partly or fully as process heat in sugar mills, in cogeneration. The sugar industry has traditionally been practising incidental cogeneration by using bagasse as a fuel for meeting the steam and power requirements of sugar processing. With advances in boiler and turbine technologies for the generation and use of steam at high temperatures and pressures, the sugar industry could produce electricity and steam for its own requirements and surplus electricity for sale to the grid using the same bagasse through optimum cogeneration.

Integrated hydropyrolysis and hydroconversion uses catalysts, hydrogen and heat in proven refinery hardware to cost-effectively convert forest, agricultural and sorted municipal wastes into fungible hydrocarbon transportation fuels such as petrol, kerosene and diesel together with char, fertilizer, water and CO_2. Although waste-to-energy technologies have the double benefit of generating electricity while taking care of urban waste, they face a major institutional hurdle. Waste segregation is an absolute requirement for waste-to-energy plants, but local bodies seem unable to segregate waste at the source. Because of the very high cost of facilities for sorting and separating waste, it is uneconomical for developers to operate under such circumstances. Processing unsegregated waste can also release toxins upon combustion. And halogens, alkalis and heavy metals in unsegregated waste require expensive gas cleaning and residue management, pushing up the costs.

Figure 3.2.

Broad technology progress in three scenarios, 2011–2050

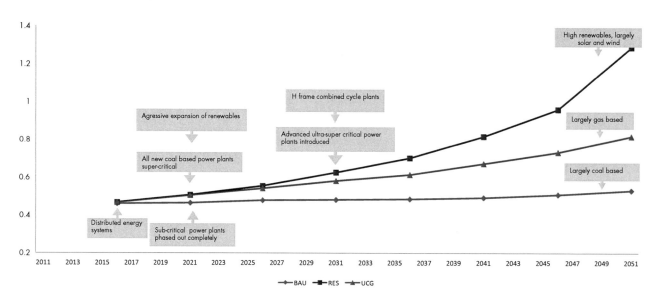

Note: Efficiency is calculated as total generation to total energy consumption. Since the supply of renewables is abundant, the conversion losses of renewables have not been considered in the calculation. As a result, very high renewable efficiency may go over 1, as in RES.

Transmission and distribution

Large capacity transformers are provided at a point, and the connections to each load are extended through long low voltage lines. The long lines results in high technical losses and low voltage distribution to the consumers. By reconfiguring the existing network to high voltage distribution system (HVDS), long length mains may be converted into 11 kV mains, and the appropriate capacity distribution transformer can be installed as close as possible to the end-user. By converting the lines to HVDS, the current flowing through the lines may come down to almost a 30th, reducing the technical losses drastically. Energy-efficient transformers, advanced metering technologies and reactive power correction near load centres can bring down losses further.

Demand technologies

Demand-side interventions could also result in significant savings.

With aggressive shifts to efficient appliances, lighting systems and efficient buildings, etc., energy demand could reduce by around 26% by 2050, compared with a reference scenarios where these technologies have limited or no deployment. The transport sector is a large consumer of energy, particularly oil, and is expected to demand much more energy in the future. To move from the heavily import-dependent crude oil technologies requires alternative fuels, with plug-in electric and hybrid vehicles two such options. Two wheelers, about 70% of India's vehicle population, are mostly used to travel short distances. Electric two wheelers that have average range of up to 140 km on a charge could be a suitable substitute. Hybrid cars, like the Nissan leaf and the Toyota Prius, combine an internal combustion engine and one or more electric motors, which push their fuel efficiency around 70% higher than that of petrol or diesel counterparts.[23] Electric and hybrid vehicles could thus bring down the energy demand for crude oil substantially.

Algal biofuels or third-generation biofuels can also help augment demand for oil. First and second generation biofuels use biomass that competes with food crops and thus are subjects of much debate. Microalgae, by contrast, can be grown on degraded and non-arable land and thus do not compete directly with food crops. They can also be grown on land with saline water. The potential for biodiesel production from microalgae is 15 to 300 times more than that for traditional crops and has very short harvesting cycles, allowing multiple or continuous harvesting. Although the technology is currently in R&D phase, algae are projected to become commercially viable in some OECD countries around 2030 for some applications, such as biojet fuels. And economies of scale may foster their implementation in India after that. The transport sector indicates an energy saving potential of 30% over the reference scenario by 2050 and these technologies could lead the way.

Technologies which may use waste to reduce raw material and/or energy consumption also need to be integrated into the sustainable solutions space to the extent possible for enhancing sustainability. Apart from focusing on waste to energy technologies, material recycling can increase energy efficiency in industry, particularly in the iron & steel and cement sectors. TERI analysis estimates that by enhancing the share of scrap-based steel production using the electric arc furnace route, energy efficiency in the iron and steel sector could improve by around 20%.[24] Similarly, energy demand in the cement sector could be reduced by nearly 30% from the reference trajectory by 2050, by enhancing the share of blended cements (portland slag cement and portland pozzolana cement). While PPC uses pozzolanic material such as fly ash from coal-based power plants, PSC uses blast furnace slag. Such technological shifts while bringing in environmental sustainability would also bring in monetary benefits to the industry units due to the energy savings and should therefore be easy to upscale if supported and backed by appropriate policies and regulatory mechanisms.

These technology interventions could reduce demand by 22% over the BAU trajectory (Figure 3.3).

Figure 3.3.

Demand technologies and energy trajectories under three scenarios, 2011–2050

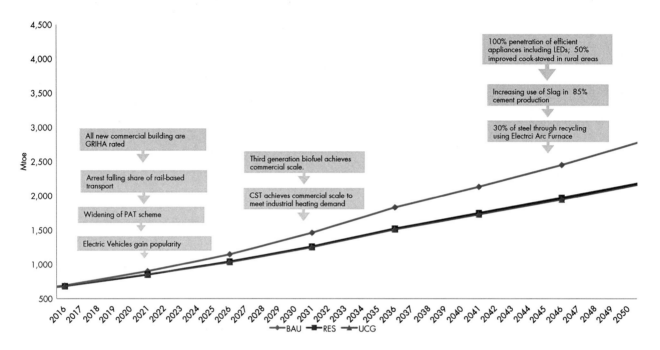

Conclusions

Given India's future energy demand, it will have to exercise all possible technology choices both on the supply and demand sides. There is no silver bullet that could easily solve its energy problems, so it needs to invest in a basket of diverse technologies and deploy them simultaneously.

The government has announced plans to increase domestic production of coal to 1,000 MT a year by 2021 and to install 100GW of solar power. Will India need both, given the energy demand in 2021? If not, what happens to the investments, and how will their costs be recovered? To avoid stranding assets, India needs to decide whether to develop through a clean or BAU trajectory. And since India is poised to make huge greenfield investments to meet its growing energy needs, it needs to make some of these decisions rather quickly. The life span of technologies is 20 to 30 years while that of infrastructure is even longer. Investment in the next few years will remain in place until at least 2050.

Planning for appropriate skill sets to work on developing, exploiting and maintaining appropriate technologies across the energy sector is another important aspect with regard to technological planning. While the Make in

India initiative is likely to spur technological development in various spheres, aligning education systems to prepare the coming generations with the requisite skill sets, and focusing on technological collaborations with advanced nations is an important ingredient for India's sustainable energy development.

There already is significant knowhow in the western countries on several transformational technologies. India has for example not only adopted the existing CFL and LED-based lighting systems to a significant extent but has also been working at adapting and improving efficiencies of these technologies and working out financing models to reduce the costs to the consumers further to enable their scale-up. Similarly, it is important that India focuses on renewable technologies, particularly solar, by not only adopting solar technologies that are already available, but working on further innovations to reduce costs and enhance efficiencies so as to enable a rapid transition. Clean coal technologies, like Integrated Gasification Combined Cycles, are another example where India can leapfrog into adoption with significant investments in R&D. But some R&D will still be needed to adapt several of the technologies to Indian conditions.

Similarly, innovation on technologies suited to the needs of the rural poor—especially those relying on traditional fuels for cooking and those that are still not connected to the grid is critical. Innovative cooking and lighting solutions could play an important role. Improved cook-stoves use a third of the biomass required in traditional cook-stoves and cause negligible indoor air pollution. While LPG penetration may be expected to increase in the future it is not expected to be taken by 100% of the population, so improved cookstoves are an ideal solution. Off-grid solar technologies are another technology that could help bring modern energy to remote parts of the country. Solar lanterns and solar home lighting could reach remote places and have the added benefit of not suffering transmission losses. Appropriate market mechanisms would be desirable for the propagation of such technologies, which are already mature and financially viable but have low implementation levels due to low awareness, little knowledge about maintenance or inappropriate incentives.

CHAPTER 4

Pricing for an Efficient Energy System

In all countries energy pricing is difficult, complex and controversial. India is no exception. The core challenge is to reconcile economic rationality (sometimes called "efficiency pricing" and the subject of this chapter) with political rationality, which is largely a matter of perceived fairness (covered in Chapter 6). Three important economic principles govern efficiency pricing and apply equally to the energy sector:

- Where a good can be traded as an import or export in a liberalized and competitive market (a "tradable good" in economic jargon), its domestic producer price should reflect the international price, adjusted for transportation costs. Pricing a domestic resource at its global price allows economic actors (and by extension nations) to assess value in terms of other tradable goods. For an intermediate good such as energy, pricing at or close to global prices also ensures that production by final users remains competitive.
- Meeting distributional or equity objectives is normally more efficient and less distortionary through income transfers than through subsidies embedded in product prices.
- Domestic indirect taxes and import duties are appropriate tools to raise revenue and to compensate for negative externalities, such as congestion or pollution. Pursuing such goals is affected by the price elasticity: a low price elasticity generates revenue without affecting behaviour, while a high price elasticity affects behaviour without necessarily generating much revenue.

Applying these principles to energy is difficult, for several reasons:

- Global markets for energy are volatile and imperfectly competitive, making an undistorted global price hard to identify.
- As with all exhaustible resources, all but the marginal producer enjoy an element of economic rent. With domestic resources, political and economic judgments determine how much of this rent to tax safely while preserving incentives for investment. But the matter is complicated when, as in India and most jurisdictions outside the United States, subsoil mineral resources are seen as national patrimony rather than private property.
- Taxing externalities, while desirable, may affect the competitiveness of end-user sectors, which may not always be compensated through either tax drawbacks or movements in the real exchange rate. (Countries differ widely in their willingness to use taxes to offset externalities associated with various forms of fuel: the United States and Europe provide contrasting examples, in levels and forms of taxation.)
- Political structures also matter, with federal systems (such as India and the United States) facing more complexities than unitary states.

Whatever the polity, it is important to think of energy pricing not only in an integrated fashion across the energy system as a whole but also for the impact on society and the environment. With energy security a growing concern in India, prices need to be high enough to ensure investment in domestic production and infrastructure. And competition from imports must remain a credible threat to force competitive pressure on domestic producers.

In energy pricing one should distinguish between the concepts of price level and price formation. The price level is the amount a customer pays in absolute terms: it needs to be set at a level that covers costs and provides a margin to induce investments and ensure the long-term sustainability of the energy system. Price formation is mediated through mechanisms to calculate the price. These mechanisms vary across the globe but are normally market-based (supply and demand of the commodity), or are linked to a substitute fuel (oil indexation in gas contracts) or are regulated to ensure a "fair" return. A market-based mechanism is usually considered the best way to get energy delivered most cost-effectively to end-users.

Why is efficient pricing important?

Energy prices need to encourage the right amount of energy to be produced domestically and/or imported (usually both) to satisfy demand. Otherwise, supply shortfalls can arise, leading to power cuts and fuel rationing, dragging down economic growth. So any pricing mechanism needs to ensure the allocation of energy to the best productive uses in the country, to ensure the maximum growth.

Rather than intervening to set prices to limit excessive profits among energy suppliers, governments often use taxes, which are more efficient. This allows them to capture surplus profits, which can then be recycled into productive social infrastructure (health, education and so on), further supporting economic growth and the population's living standards—creating a virtuous circle.

A rational approach to energy pricing also improves the efficiency of energy use and mitigates the environmental impacts of energy consumption, often by sending the right signals to "nudge" consumers to select more sustainable forms of energy or purchase more energy efficient devices.

It is also important to price in other policy goals, such as minimizing environmental impacts—or increasing sustainability, ensuring adequate energy supply and maximizing industrial competitiveness. These benefits sometimes run in opposite directions, requiring holistic pricing as well as policymaking to avoid unintentional consequences (further discussed in Chapter 6).

India a market in transition—the need for efficient energy markets

The energy sector in India is characterized by monopoly players, state-run corporations, controlled pricing and high barriers to market entry. The lack of transparent price signals, the presence of controls on supply (such as the Natural Gas Utilization Policy) and other market barriers have reduced incentives to invest, improve efficiency or rationalize supply and consumption.

High barriers to entry in the energy sector arise from legislative barriers, uncertain output, high capital and technological requirements and the need to achieve economies of scale. And because energy contributes to eradicating poverty (as recognized by the United Nations in the Sustainable Development Goals) and affects standards of living, governments are inclined to keep energy prices low, usually through controls or high subsidies for fossil fuels. They also attempt to avoid fluctuating and high prices for electoral reasons. Such constraints muddy the transparency of pricing in certain fuels and lead to high under recoveries by energy companies.

The Indian government has allowed pricing structures to evolve to some degree, introducing private sector competition over the past few decades. But it still heavily controls several functions. Inefficient extraction techniques, irrational supply linkages, inefficient consumption and wasted energy are still widespread, and provide considerable scope to approach global best practice.

Historically the government has controlled prices of key petroleum products to cushion the public from price fluctuations and ensure affordability, but this has increased its subsidy burden (Figure 4.1). While kerosene and liquefied petroleum gas (LPG) have remained subsidized (with the intent of improving access for the poor), petrol and diesel prices were decontrolled in 2010 and 2014 (Figure 4.2), after a period of reducing under-recoveries through price increases from 2013.

Figure 4.1.

Petroleum subsidy costs in India

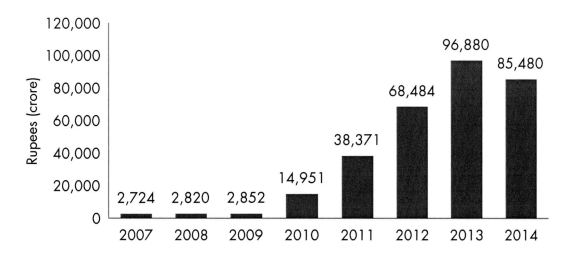

Source: MoPNG 2014.

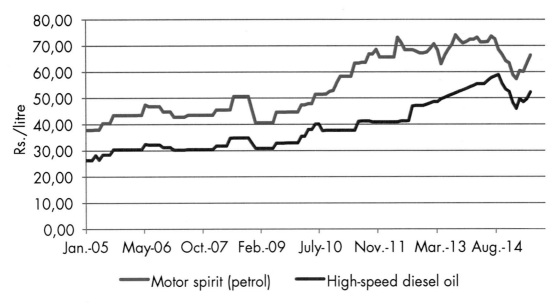

Figure 4.2.
Retail petrol and diesel prices in Delhi

Source: PPAC n.d.

Energy subsidies can sometimes help manage affordability and accessibility especially in developing markets like India, but they can also create perverse environmental, fiscal, macroeconomic and social consequences. Subsidies are also highly inefficient in supporting low-income households, since richer households typically capture most of the benefits. Today, however, the current low-energy price environment and the continuing fiscal pressures on the Indian government offer opportunities for further reform beyond recent progress.

Efficiency pricing to integrate domestic and international energy supply

India faces formidable challenges in meeting its energy needs. As demonstrated by the scenarios in Chapter 1, although it will be possible to boost indigenous coal and hydrocarbon production, any increase will not be enough to satisfy its growing energy demand. India will therefore remain a net importer of oil, gas and coal for the foreseeable future. In fact, as needs grow, it will become a larger and more important player on the international energy market (covered in Chapter 5).

The turbulence of geopolitics continues to overshadow global energy prices, worldwide energy supply and demand, and domestic pricing policy. Multilateral trade agreements have not really ensured a worldwide free market in energy. Globally, prices and pricing are often driven by concerns of energy security, regional hegemonies and other externalities, like the impact of technology or climate change (Box 4.1).

Pricing externalities

In economics an externality is the cost or benefit that affects a party who did not choose to incur that cost or benefit. For example, if the negative effects of air pollution from coal-fired power plants begin to outweigh the positive effects of cheap and reliable electricity, citizens will start to resist further coal-fired plants or demand that existing plants be upgraded with pollution-control mechanisms.

The need for such corrective action was recognized a long time ago by the British economist, Arthur Pigou, who first noted that if one taxes the activity generating a negative externality, the party responsible would reduce that activity's intensity. And by selecting the right tax level, the authorities could achieve the goal they want in reducing negative external effects.

Coal pricing in India

Even though Coal India Limited (CIL) was given nominal autonomy for setting prices in 2000, the government makes the strategic decisions. Coal-producing companies still cannot freely negotiate prices with their customers, and vice versa.

India has two types of prices for coal. Government-approved "notified prices" apply to all consumers that have signed a fuel supply agreement with CIL and Singareni Collieries Company Ltd. (SCCL). The power and fertilizer sectors have priority in buying coal at these prices. The second type is set by an electronic auction (e-auction), where coal companies are allowed to fix a "reserve price" that is generally 20–25% higher than the notified price. Coal e-auctions conducted by CIL and SCCL show that consumers are willing to pay a premium over the reserve price.

The notified price for coal has the effect of distorting market signals and is one of the reasons that capital allocation to new mining projects has been insufficient, contributing to the growing shortfall in output

Coal prices will need to rise to ensure investment

Despite large coal reserves India struggles to meet its coal demand from domestic supplies and so is forced to import. Although many reasons explain the need for imports, including land disputes and difficulties in obtaining permits, price levels are also key. And to the extent India decides to rely on ever deeper and more expensive underground coal deposits, production could benefit from more modern mining technologies. These will demand capital, in turn requiring a better investment climate, including prices that will reward for the outlays.

Beyond augmenting domestic coal production, India will need to encourage further coal imports to meet its energy needs, making it a major force on the international coal market in view of the import volumes required. Indonesia has already become the principal thermal coal supplier to India because of its low-cost, high calorific value coal and its proximity to India.

In the past China was the largest thermal coal importer from Indonesia, with India a close second. But the Chinese government is contemplating restricting lignite imports on environmental grounds and switching to more environmentally friendly gas. This means that India is likely to become Indonesia's biggest coal market, and so will set the price for Indonesia's coal, moving it from a price taker to a price setter.

Longer term prices should be set by the market

Greater private participation in captive and commercial mining will over the long run yield benefits to the sector and the wider economy. Private and foreign mining firms could employ industrial best practices and technologies in India, improving productivity and operational efficiency. Coal price discovery in commercial mining, without any end-use restrictions, would incentivize foreign investment from global mining companies and from equipment and technology firms. This pricing approach would also help create an efficient coal market—triggering productivity gains in currently uneconomic mines or idling too costly operations—thereby increasing coal supplies at competitive prices in India.

Such liberalizing moves require independent regulation to improve the exploration and allocation of resources, to regulate e-auction and notified coal prices and to help promote a competitive coal market. The other energy markets (gas, oil and electricity) would also benefit from independent regulation, resulting in similar improvements as just highlighted for coal. Box 4.2 and Chapter 6 expand on the need for an independent regulator.

The role of government in energy and the need for an independent regulator

Wherever possible, energy markets—not just that for coal—should be competitive, where pricing and resource allocation are determined by market forces, with the minimum of government intervention. This is why the oil, gas and power sectors in India have been progressively liberalized. Yet successive Indian government have felt compelled to intervene in price-setting and resource allocation, often to protect such economic sectors as fertilizers and such populations as the poor. Investment in infrastructure and supply has consequently suffered, causing energy shortages.

The persistence of energy market distortions and the resultant costs to major Indian energy companies (which carry part of the subsidy burden) put pressure on energy firms' credit ratings and raise their cost of capital — and ultimately the cost of energy supply in the long term. Direct government intervention can also unnerve global—and local—investors, as it typically increases market uncertainties and expected economic returns. This makes companies more reluctant to invest and even induces them to pull capital out of the country.

Conversely, competition alone has been shown around the world to have limitations, which can best be managed by independent regulation. A regulator helps ensure both that market players do not abuse any position of market dominance or natural monopoly, and that they adopt competitive market behaviour. This means that market players do not excessively overcharge end-users. As energy markets evolve, most move to an independent regulator model, one in which the government sets policy, but an independent regulator is responsible for day-to-day market oversight. Such independence also makes regulation more predictable and less prone to arbitrary decision-making.

A shift to such a model would be an important next step for the Indian energy market as it continues to evolve in size and sophistication. It would also no doubt increase investor confidence—especially if the regulator is demonstrably independent from government, has transparent decision-making process and demonstrably bases decisions on rational economic analysis.

Gas pricing

Before November 2014 the government in India set gas prices, which private companies investing in upstream exploration and production (E&P) regarded as too low, especially for complex offshore gas field developments. The consequences of such policies included shortages of domestic gas, losses of state revenues, wasteful use of natural gas by users and demand inflated by domestic industries that rely on unsustainably cheap gas.

In recent years domestic supplies have not just failed to keep up with demand, but have fallen, leading buyers to seek higher-priced LNG on the global market. India has a long-standing policy of capping the price of domestically produced gas, as well as that of the end-user, weakening incentives for investment in domestic gas E&P. Further, the number of wells drilled in India's offshore gas fields has declined since 2007, revealing a growing reliance on Indian state-owned firms as private and foreign entities looked elsewhere. This capital reversal has removed a potential source of growth from the economy. Limited revenue from gas sales has further impeded investment in new infrastructure, most notably gas pipelines. As a result, power generators have been forced with increasing frequency to leave gas fired generation assets idle. Particularly after 2010, gas supply constraints forced a sharp decline in gas-fired power plant utilization rates across the country. The price caps have also weakened incentives for end-users to find efficiency savings, leading to waste and stimulating additional consumption, exacerbating supply shortfalls.

Another distortion that affects pricing and thus revenue generation is The Natural Gas Utilization Policy, covered in Chapter 6.

Gas pricing reform and impact

In November 2014 the government adopted an unprecedented reform of domestic gas pricing. It linked the domestic gas price to a volume-weighted formula of several world gas price markers—US Henry Hub, UK National Balancing Point, Canadian Alberta Gas Reference price and the Russian domestic gas price. This move is to be welcomed because it introduced a formal transparent price mechanism. But the prices in the formula are underpinned by the fundamentals of the markets in countries or regions they represent, which are likely to change independently of India's.

However, India's new gas pricing formula has failed to incentivize upstream producers to step up investments. This means that operators like Reliance and Gujarat State Petroleum Corporation are deferring investment in key offshore projects since current prices make their projects unviable. This is also illustrated by a review of existing Indian gas reserves by the Oxford Institute for Energy Studies, which shows that at current or marginally higher price levels, the reform is unlikely to reverse the recent decline in domestic gas production before 2020.

In March 2015 the Cabinet Committee on Economic Affairs approved price pooling for domestically produced and imported natural gas to provide domestic manufacturers with better price signals.

Further reform still needed

Despite the reform India still lacks a full price-formation mechanism that reflects the national energy market. In China's reform, gas prices have been determined by the fuels they replace in the domestic economy (Box 4.3). A next step for India could be to link domestic gas prices to a discounted average price of fuel oil, unsubsidized LPG, naphtha and distillates on the domestic market. This would increase gas prices and so incentivize gas production, but as with China would reduce overall energy costs, due to substitution by gas. But akin to the recent completion of product price reforms for petroleum in India, those for gas should be easier to carry out in the current low global energy price environment.

A higher gas price is not the only variable that affects upstream exploration and production. The situation can also be altered by a change in the upstream fiscal terms or a reduction in the perceived risk associated with an upstream investment in India.

Box 4.3

A global perspective—China's gas price reform

China's low regulated gas prices have stimulated demand, but over the past decade growth in domestic supply has struggled to keep pace with rising demand, particularly in the economically red-hot coastal provinces. To its credit the government has recognized that regulated prices inhibit overall gas development and has in response increased gas prices regularly through a series of price reforms.

As oil-indexed LNG imports have risen to meet rising gas demand, China has sought to develop domestic unconventional gas resources, for security of supply and to keep down costs. Industry also recognized the importance of reforming the domestic gas pricing system to increase domestic production.

From July 2013 city-gate gas prices were divided into two categories: existing demand (based on 2012 contracted onshore and imported pipeline volumes) and incremental demand (based on incremental supply). This allowed for differential pricing between the demand categories, helping manage the transition to higher gas supply prices and limiting the price impact on end-users.

Another change was a shift from a cost-plus system for onshore production to an oil-linked price structure on a netback basis to oil-indexed city gas prices. This enabled gas prices to rise, creating incentives to boost production. But because gas was displacing more expensive oil products (LPG and fuel oil), the overall energy costs were reduced.

This dynamic price formation mechanism has since been modified to reflect regional market fundamentals as the reforms have progressed and as China's gas market expanded across provinces.

Electricity pricing

After the National Tariff Policy was introduced most utilities started adopting time-of-use tariffs for large consumers. The price differentials are, however, determined somewhat arbitrarily by the regulatory commission to provide overall revenue recovery for the utility. A large part of the revenue is used for cross-subsidizing smaller consumers, instead of reflecting the cost of procuring more expensive power to avoid load shedding. The elevated end-user industrial electricity tariffs, as well as unreliable supply, leads to many industrial consumers to produce their own electricity. This capacity is often coal-fired, relatively inefficient and under-used.

All this has produced financial problems for the local distribution companies, often worsened by shortfalls in compensation due from state governments. This is further compounded by poor metering and inefficient billing and collection. Because of these financial problems, the local distribution companies cannot invest as much as they should to upgrade ageing and high-loss parts of the network.

India's distribution utilities are unwilling to purchase higher priced electricity from new electricity plants, which run on imported coal or LNG, and instead resort to load shedding. They also have difficulties meeting obligations to purchase power from renewable energy sources. The outcomes are power cuts in periods of high demand, underuse of power plants and market distortions. Peak power tariffs should always reflect the cost of purchasing and producing peak power.

In addition, electricity prices that do not reflect the cost of production and investment, or that lead to underuse of power stations, will have a negative impact on the returns on the investments made. This ultimately damages the investment climate, leading to deferred investment decisions, rising electricity supply shortages and the increased risk of blackouts.

The solution to India's electricity problems is not only to raise average tariffs to cover the cost of production, but also to deal with the inefficiencies and bottlenecks. Power tariffs in theory are regulated by State Electricity Regulatory Commissions, but state governments often intervene to prevent increases. As seen, the central government keeps tariffs low by managing the cost of fuel (coal and gas), which has the effect of lowering investment in these sectors, leading to further supply shortages.

Raising energy prices (as only one means) to ensure investment

The key primary fuels for India will remain coal, oil and gas, with a growing role for renewables. (India's dependence on imported fossil fuels rose to 38% in 2012 according to the US Energy Information Administration, even though it has large domestic fossil fuel resources.) The first three are all imported at the margin, with India also a heavy exporter of oil products. Because imports are fairly lightly subject to import duties, the main pricing issues relate to domestic E&P of the equivalent domestic resource, and passing price signals down the value chain to the end-user, whether bulk or retail. The dichotomy of "higher primary energy pricing to attract investment" and "low pricing of secondary energy to ensure better access" has also created major distortions in the energy market in India, although for many industrial consumers, energy prices are very high in real terms, undercutting their global competitiveness.

Energy prices in India therefore need to be set at a level that encourages investment in supply and infrastructure, coupled with targeted help to the poor to ensure affordability and energy access in a move away from subsidies. Raising prices to international levels will also maximize domestic supply, reduce imports, incentivize more efficient use of energy and provide an additional source of revenue for the government through taxation. But these moves will also require meaningful pricing reforms, especially in energy distribution.

Managing the transition to higher prices

Prices need to be raised to a level that will ensure investments not only in the power sector but also in other energy supply sectors. This level will need to at least cover the cost of fuel and capital—and beyond that, provide a return to the investor. For domestic energy supply, efficient pricing needs to be transparent (while mitigating price shocks) to help manage price volatility and to deliver long-term use of sustainable resources at the lowest cost possible. The total cost (fuel, investment and so on) and a reasonable profit (to ensure investment) should be spread across all suppliers of energy to the end-users.

As a general rule, all primary energy sources should be priced at trade parity at the point of sale—that is, at the free-on-board price for products for which India is a net exporter, and the cost, insurance, freight price for which it is a net importer.

To cushion domestic energy prices against short-term volatility of prices on the international market, domestic prices can be set on the basis of median prices over the previous one, three or six months. This will smooth the impact of price shocks and help manage the transition to higher prices. Another way to manage volatility is through financial instruments, allowing the forward price of a commodity to be locked in, a process known as hedging. Normally these financial products are available in mature and deregulated markets like the United Kingdom and the United States, which means it could be an option for India in the longer term. Stocks and storage can also help manage short-term supply interruptions, though they require investment and thus come at a cost to the overall energy system.

Reforming energy subsidies

Reform to energy subsidies is high on the global international policy agenda, reflecting the increasing recognition of their perverse fiscal, macroeconomic, social and environmental impacts. They impose large fiscal costs, which need to be financed by increasing public debt or taxes, and they can often crowd out the more productive public spending (education and infrastructure) that makes economies more competitive globally and improves their economic growth prospects.

Energy subsides also discourage investment in energy efficiency, since they often make it uneconomic to invest in more expensive but more energy-efficient technologies and products, and increase energy demand to unsustainable levels, leading to supply shortages and wasteful energy use.

Total energy subsidies in India amounted to some US$47 billion in 2013, equivalent to 2.5% of GDP. Most were for oil and associated oil products (Table 4.1). Energy subsidies are also highly inefficient at providing support to low-income households, as most of the benefits usually go to richer households. Targeted financial support to the poor is usually the more equitable option.

Table 4.1
Energy price subsidies in India (bln USD 2013)

Fuel	2011	2012	2013
Oil	33.4	39.2	36.6
Gas	3.4	3.1	4.2
Coal	0	0	0
Electricity	3.6	4.4	6.2
Total	40.4	46.7	47.0

Source: International Energy Agency Fossil Fuel Subsidy Database.

In natural gas price reforms specifically, one of the primary concerns of policymakers is that higher gas prices will translate into higher fertilizer subsidies. These subsidies could, though, be completely offset by the increase in government revenues as gas production is valued at a higher price. Any increase in gas prices above current levels would reduce the absolute size of government subsidies through higher royalties and taxes.

Subsidies on gasoline and diesel have already seen progress. Gasoline prices were deregulated at retail level in June 2010. In October 2014 India's Union Cabinet Committee on Economic Affairs issued a directive to liberalize diesel prices for all consumers, which concluded two years of regulated price increases for retail diesel prices by a monthly maximum of Rs. 0.5 per litre. These price adjustments have gradually closed the gap between India's subsidized diesel prices and international prices.

Unlike diesel and gasoline, reform of LPG and kerosene pricing has been much slower, with the government still controlling the price to residential, commercial and agricultural end-users as part of its wealth redistribution and poverty alleviation policy, reflecting the role of these fuels in providing lighting and cooking fuels to the poorest segments of society. But the distribution channels are highly ineffective, allowing leakages to other sectors. The government has begun to restrict the quota of subsidized fuels that each household can receive, and the government is committed to make them more efficient. A nationwide implementation of the Direct Benefit Transfer Scheme for LPG (DBTL) to move the LPG sale to market prices (and directly transferring subsidy to beneficiary bank accounts) and reduce diversion was carried out in early 2015. A CEEW study, based on a primary survey of more than 1,200 households, 90 LPG distributors and interviews with implementation officers, bank managers and senior members of the oil marketing companies, puts forward the learning and experiences from this large-scale implementation. The analysis indicates that strong political will and leadership, juxtaposed with institutional coordination and intensive advertising and communication campaigns, were the key drivers behind the scheme's successful implementation. The study found high levels of consumer satisfaction and a positive outlook of other key stakeholders towards the scheme. This is a stepping stone for future reform in targeting subsidies better.

The government also launched a "Give it up" campaign to encourage the wealthiest consumers to abandon their LPG subsidies. By February 2016 more than seven million households had voluntarily given up their LPG subsidy.

Moving to renewables

Although India may be keen to increase renewable generation for security of supply, the heavy share of such generation in Germany (Box 4.4) has itself led to security of supply issues in electricity. India could face this issue if its generation mix has too high a share of renewables. And renewables on a large scale could result in large subsidies, as in Germany, which would be a major drain on government finances and likely lead to higher electricity prices for end-users.

Box 4.4

A global perspective—the German electricity market and subsidies for renewables

Spurred on by generous government incentives, Germany has seen unprecedented growth in renewable capacity in recent years and in non-thermal renewable sources (wind and solar photovoltaic), which account for almost 40% of power generation capacity. Such growth has come at a heavy cost to consumers.

The levy imposed on domestic consumers' electricity bills to cover the cost of renewable subsidies has risen from €0.25 cent/kWh in 2000 to over €6.4 cent/kWh in 2014, and now makes up almost 20% of a typical household's electricity bill. The cost of subsiding renewables through this levy in 2014 was €21.7 billion.

Unlike end-user prices, wholesale power prices have been under pressure, prompting some operators to mothball gas-fired and coal plants. The problem is that such plants are needed as back-up to intermittent renewable generation like solar and wind, threatening system integrity on the grid. In an effort to maintain the competitiveness of German industry, the government has all but exempted energy-intensive industrial consumers from paying the levy for renewables.

In a nutshell: policy has been very successful in delivering a large amount of renewable generation capacity. But it has been very expensive, largely exempts German industry and affects security of supply, which may well require getting the gas-fired generators on the grid again to provide backup capacity—at an additional cost.

Energy conservation and innovative pricing

Good for the environment, energy conservation also helps to lower overall energy demand and so reduces the investments required. But it works only when end-users invest in higher efficiency appliances, which are normally more costly than standard equipment. And energy prices have to be set to incentivize these investments, because if energy prices are too low, they increase fuel consumption.

Most energy conservation measures require upfront investments, deterring many low-income families from buying more expensive appliances. Innovative pricing can help deliver energy at lower cost to end-users by reducing the overall system cost and by easing the initial outlays. It moves from higher cost to lower cost products over the longer run, also bringing environmental and health benefits.

The Lighting a Billion Lives programme, led by the India Energy and Resource Institute, replaces kerosene lamps with solar lanterns. Micro enterprises rent out high-quality and cost-effective solar lamps for a low fee in unelectrified or poorly electrified villages. Since the users rent the lamps rather than buy them upfront, solar lighting is affordable for poorer sections of society. Indeed, the rental fee is less than previous outlays for kerosene.

This innovative pricing mechanism also demonstrates that thinking holistically and in an integrated way, provides not only a more environmental beneficial outcome but leads to an reduction in total system energy costs.

Taxation

Taxation and pricing often go hand in hand, in terms of the total price of a commodity paid for by the end customers, the effect on the rate of return on energy investments and so on investment decisions, plus their use to achieving various policy objections, for instance minimizing environmental impacts.

So Central and state taxes on energy supplies need to produce optimal fuel choices and investment decisions. If taxes and subsidies are not equivalent across fuels, then the relative prices of fuels can be distorted. The equivalence should be based on energy content, often referred to as "calorific value."

Environmental taxes and subsidies can however be levied to affect fuel choices. These differential taxes can be justified if they reflect environmental externalities, the so called "polluter-pays" principle or "consumer-pays" principle. But these should always be consistently applied to attain the necessary environmental objectives at the least cost.

Upstream hydrocarbon taxation

Upstream taxation is one of three key components necessary for upstream investment. The other two are the price of the commodity sold and the perceived risk of an upstream investment. Any upstream fiscal regime should incentivize the maximum possible amount of upstream investment, while trying to capture as much rent as possible, without affecting investment levels. It should also ensure the correct balance of risk and reward for both the investor and the government. Where there is a history of constant government interference, or where the government constantly shifts the risk balance in its favour, the perceived risk of investing in upstream E&P activities will be higher.

The proper design of upstream fiscal models—including specific terms, allowances and tax levels—can ensure that the government has a consistent policy for a few decades, in order to minimize uncertainty and create a positive investment climate.

Many upstream fiscal regimes apply specific allowances or rates for more expensive projects, to ensure the necessary incentives are in place to maximize total investment and so production within a country. For instance - deepwater development tend to cost more than shallow water or onshore developments, and therefore deepwater developments normally attract additional allowances and incentives.

Specific allowances can also be introduced to encourage exploration, important given India's underexplored hydrocarbon resources. And specific incentives and measures can target the development of small or marginal upstream fields—or fields that are more technically difficult (increased risk) or more costly (lower returns). With this in mind, the Indian government approved the Marginal Field Policy in September 2015, shifting from profit sharing contracts to revenue sharing contracts (further explored in Chapter 6).

CHAPTER 5

India in Global Energy Markets

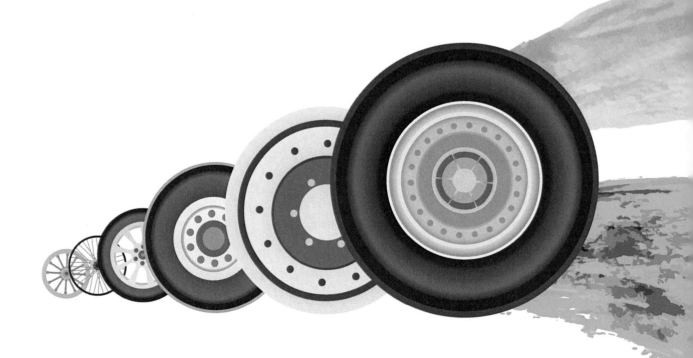

Seeking energy security, not energy independence

In 1991 a balance-of-payments crisis forced India to confront the inevitable—that a closed economy in pursuit of self-sufficiency could not reach its full potential on poverty reduction, economic growth and job creation—and its rightful place in a changing geopolitical world. Over the past 24 years India has integrated incrementally into the global economic order, sometimes at a pace that critics find painfully slow. Yet the direction has been clear: more trade, more investment, deeper integration and freer movements of goods, services, people and ideas. India is now the world's fourth largest economy (in purchasing power parity terms), with a trade-to-GDP ratio of 50%.[25] The Indian economy, certainly, is no longer autarkic.

So how will India's energy system go through a similar transition—in the terms, measures and institutions—from autarky to deeper integration into global energy trade and investment? By 2030 India's share in daily oil trade is expected to be 12.5%,[26] up from 7.4% in 2014.[27] (For comparison, the share of the United States in global oil imports was 26.4% at its peak in 2005;[28] the equivalent figure for the European countries [excluding Central Europe] was 38% in 1980.) In other words, the sheer momentum of India's economic growth, the continuing shift of people to urban areas and the imperatives of energy access will ensure that energy demand will grow so much that energy autarky will not suffice. In the 21st century India is on track to assuming the role in global energy trade that Europe had in the 20th century. It will not be the biggest energy consumer. But it will no longer be possible to play only at the margins. Indeed, India will be a "swing voter" in global energy markets with strong national interest in well-functioning markets.[29]

The Indian government of course recognizes this inevitability, no doubt with growing trepidation. It is deeply concerned about the rising share of crude oil imports. According to the modelling scenarios, this could increase from 65% of oil demand in 2000,[30] to 78% in 2015[31] and to 90% in 2030.[32] Coal imports have been rising year on year. And natural gas now accounts for 12%[33] of installed electricity capacity, in addition to being a primary fuel in fertilizer and chemical industries. Imports of gas are likely to rise to 89 billion cubic metres (BCM) by 2030, more than five times the level in 2013–2014.[34] With little investment in domestic upstream exploration, even fewer discoveries and flat production of primary energy fuels, there are concerns about how vulnerable India is becoming to energy shocks.

In response, two strategies are often recommended. The first is to aspire to energy independence. This does not imply zero imports of primary energy, but it does aim for a reduction rather than an increase in the share of imported oil, gas and coal. For instance, the government wants oil imports to fall to 67% of demand by 2022 and to 50% by 2030.[35] The second strategy has been to buy acreages in oil and gas fields and in coal mines beyond India's shores.[36] The assumption is that such overseas assets will deliver energy resources to India's shores in times of crisis. These strategies are inadequate, however, and difficult to achieve, even for large economies.

The quest for energy independence has not been India's alone. Faced with the first oil shock, President Nixon set a goal (in November 1973) to meet energy needs "without depending on any foreign energy source." This effort was further challenged with the second oil shock, the result being that President Carter declared that the Persian Gulf was of "vital interest" to the United States and would be defended "by any means necessary, including military force." More recently, the United States has had a surge in domestic production (tight oil accounted for 49% of its crude oil production in 2014), making it the largest producer of oil and gas combined. Even so, it cannot disengage from global markets, because global oil prices would still have an impact on domestic energy investment, production and consumption.

Oil is a fungible commodity (and a financial asset), so decisions in other large domestic markets have implications for global oil trade and prices. India is already more dependent on energy imports than China or the United States. With a much smaller domestic output of oil and gas and growing reliance on imported coal, it can ill afford to seek energy independence by relying only on domestic resources.

The second strategy—purchasing overseas acreages—is an extension of India's building strong bilateral relationships with major energy suppliers, often featuring long-term contracts. Relying on a few suppliers of oil, gas and coal assumed that adequate quantities of fuel would be delivered and that price fluctuations would not matter. The limits of that strategy have been exposed as India struggles to persuade OPEC countries to renegotiate long-term contracts. It wants to benefit from the fall in oil prices over the past year and to stop paying the "Asian premium," which it claims has been traditionally imposed on major Asian energy importers.[37] Saudi Arabia, in turn, has countered that there is no Asian premium.[38] And empirical studies have found mixed results.[39]

The reality is that for emerging large energy importers, the shift from long-term contracts to relying on diverse sources, taking advantage of lower spot market prices and hedging via forward contracts are all likely to be part of a new strategy for engaging more deeply with global energy markets. With a growing economy and more cash to spend, India's public sector oil and gas companies have been investing in oil and gas fields. But the needed variety in sources of supply is not guaranteed solely by owning energy assets.

For India, energy security will not be the same as energy independence. Such security will require meeting four imperatives: assured supply, safe passage, secure storage and a seat at one or more international forums in the international oil supply and trading game.

In this quest India would first have to seek an assured supply of energy resources. While ownership of assets might have a limited role in times of crisis, more effort will be needed in boosting India's diplomatic capacity and aligning it with the commercial interests of India's public and private energy companies.

Second, India would need to ensure safe passage of overseas energy supplies. This will be partly a function of India's ownership of—or access to—a shipping fleet. Compared with other major energy consumers, India's share of oil and gas tankers is low. Safe passage will also require naval capabilities for India to become a net security provider in the Indian Ocean. India has been pursuing regional as well as bilateral cooperation on maritime security in the Indian Ocean, but such engagements need to be intensified even further. Its naval assets will also have to work with other navies in protecting energy supply routes beyond the Indian Ocean, particularly in the South China Sea, from which new supplies of energy might flow in the future.

Third, India would need secure storage as well as infrastructure and management capabilities to store and transport energy resources on its territory, as a buffer for emergencies. But its storage capacity for strategic reserves, including that nearing completion and planned, is low relative to that in OECD countries or China (current and planned). In any case, capacity alone would not suffice. India can learn from members of the International Energy Agency (IEA) on their decisions for siting storage facilities (including in other countries) and, more important, on their institutional capacity to manage strategic reserves.

Fourth, perhaps the hardest question for Indian policymakers is determining which international forums could offer a platform for them to articulate their concerns about the global energy system and learn the rules and codes of conduct that govern energy relations between other major economies. Because there is no global energy regime, India would need to identify the key functions that a regional or plurilateral energy regime could perform and that would otherwise be hard to do unilaterally. These functions include assuring transparency in energy markets, cooperatively managing strategic reserves, jointly patrolling energy supply routes, arbitrating disputes and pooling resources to lower insurance premiums on transporting resources.

This chapter begins with a discussion of how countries have come to define energy security in line with their national interests and what approach ought to be appropriate for India. It then discusses the four imperatives and concludes by identifying some unknowns that India would have to be aware of as it becomes more deeply integrated into global energy markets.

Understanding energy security

Vulnerabilities in energy security were exposed during the two world wars, on all sides. If British politicians were worried about oil supplies during the first conflagration, Japan's entry into the second was in part due to the energy embargo on its economy. During the Cold War the quest to control energy supply routes continued. Even in relatively peaceful times, the two oil shocks triggered by OPEC in the 1970s made energy security a paramount purpose for the world's leading economies. But what is energy security? And what does it mean for India?

Evolving definitions of energy security—more than quantities and prices

The International Energy Agency (IEA) defines energy security as the uninterrupted availability of energy sources at an affordable price. But the pursuit for energy security varies with time. In the short run energy security focuses on the ability of the energy system to react promptly to sudden changes in the supply–demand balance. In the long run it mainly concerns making timely investments to ensure that energy supply is in line with economic developments or changing environmental needs.[40]

Several leading economies have understood energy security in different ways, depending on their circumstances. The US Energy Independence and Security Act of 2007 states its purpose as "to move the United States towards greater energy independence and security, to increase the production of clean renewable fuels, to protect consumers, to increase the efficiency of products, buildings and vehicles, to promote research on and deploy greenhouse gas capture and storage options, and to improve the energy performance of the Federal Government, and for other purposes."[41] This is a more expansive definition but other countries have evolved their thinking in similar ways. Germany's New Energy Policy (2012) borrowed from the IEA definition but also added that greenhouse gas emissions would have to be reduced while increasing the share of renewable energy and improving energy efficiency.[42]

Japan, a major importer of oil and gas, developed a Fourth Strategic Energy Plan in 2014—three years after the Fukushima nuclear accident had imposed a drastic rethinking about energy sources and uses. Japan's plan also expanded on a narrow definition of energy supply (or energy security) to include a focus on realizing low-cost energy through energy efficiency measures and on increasing efforts to ensure environmental sustainability.[43] It has therefore planned, among other targets, to double the self-sufficiency ratio by 2030, to double zero-emission power generation and to be the best in the world on industrial energy efficiency.

Other countries might not explicitly define energy security but have plans or programmes that indicate their core priorities. Under its Green Growth Strategy the Republic of Korea wishes to strengthen "energy independence," create new growth engines, and improve the quality of people's lives.[44] China's 11th Five Year Plan (2006–2010) set priorities for increasing domestic energy supplies and promoting energy conservation. Its 12th Plan went further: diversifying energy resources and import routes, enhancing existing sources and exploring and developing new domestic resource locations.[45]

The common thread across all these definitions is that countries are not satisfied with a definition of energy security limited only to quantities and prices. Developed and fast-emerging economies have seen value in diversification (on the supply side), efficiency and conservation (on the demand side) and sustainability (on the environmental front).

Energy security for India

For India the problem was different until recently. Its energy supply was limited relative to demand (and still is), but overseas energy security was not the primary concern. Domestically produced coal was the mainstay of the modern energy sector, for fuelling electricity or large industry. India was, of course, dependent on imports of oil (and to a much smaller extent, gas, which was first imported in 2004). Major episodes in West Asia have affected India's economy (Table 5.1). Among them, the 1979 oil shock triggered a decade of macroeconomic pressure, with oil price fluctuations combining with rising debt service obligations. The 1990–1991 Gulf crisis imposed severe pressure on the balance of payments, eventually resulting in the rupee devaluation in July 1991 and the crisis-driven onset of economic reforms.[46] The US cruise missile strike on Iraq in 1996 again led to crude supply disruptions in India, in turn tripling its current account deficit. India's energy-related crises, in other words, had clear macroeconomic implications for a small and vulnerable economy dependent on oil imports. This was different from pure quantity-focused supply-related security concerns.

Table 5.1.

Major supply disruptions facing India since 1979

Year	Crisis	Impact on global oil	Impact on India
1978–1979	Iranian revolution	Oil disruption ~5.6 mbd; hike in spot prices; artificial demand for oil	Devaluation of the rupee
1980–1981	Early Iran–Iraq war	Oil disruption ~4.1 mbd; steep price hike	India faced shortage of crude oil and oil products
1990–1991	Gulf War I	UN embargo on oil from Iraq (4.3 mbd) followed by steep price hike	India was forced to buy 4.75 million tons of expensive oil in the spot market; value of oil imports increased by 0.6% of GDP, contributing to half the increase in the BoP gap
1996	Kurdish civil war and US cruise missile strike on Iraq	Iraqi oil didn't enter market; prices shot up	India's current account deficit almost tripled
2003	Gulf War II	Price hike and supply disruptions of ~2.3 mbd	Domestic fuel prices had to be increased
2010	Iranian oil embargo		Forced to reduce cheap imports from Iran by ~38% over 2010–2013

Source: Robert McNally (2012), "Managing Oil Market Disruption in a Confrontation with Iran," Energy Brief, Council on Foreign Relations, http://www.cfr.org/iran/managing-oil-market-disruption-confrontation-iran/p27171; Gulshan Dietl (2004), "New Threats to Oil and Gas in West Asia: Issues in India's Energy Security," Strategic Analysis, IDSA. Available at http://www.idsa.in/strategicanalysis/NewThreatstoOilandGasinWestAsiaIssuesinIndiasEnergySecurity_gdeitl_0704.html
Arvind Virmani (2001) "India's 1990–91 Crisis: Reforms, Myths and Paradoxes," Planning Commission, available at www.planningcommission.gov.in/reports/wrkpapers/wp_cris9091.pdf

India's response to such price volatility has lain in securing long-term contracts with a few key oil exporters, but that approach is becoming less tenable. Oil (and increasingly gas and coal) are no longer merely macroeconomic headaches for India. With a much larger economy now, India's energy security is also intimately linked to actual barrels of oil equivalent being available to sustain industrial and overall economic growth. This lesson was brought home during the sanctions on Iran: India was forced to reduce cheap imports from that country by 38% (Box 5.1).

Box 5.1.

Iranian sanctions and India's energy security challenges

In the past decade Iran has been subjected to three rounds of sanctions: by the UN in 2006, the United States in 2007 (tightened in 2010) and the European Union in 2012.[47] US sanctions in particular made India's trade with Iran difficult. (Iran was India's second largest oil supplier, accounting for 11% of total crude imports in 2010.) As a result, India faced three obstacles: reducing imports from Iran, paying Iran and securing maritime (re) insurance. The first challenge was handled by gradual reductions in annual imports from Iran, sufficient to exhibit considerable reductions to obtain US waivers for continued imports[48]: such imports dropped by 38% from 17.2 MMT in 2010 to 10.7 MMT in 2013, while their share in total crude imports fell by half from 11% to 5.5%.[49]

India attempted to meet its energy needs in the medium term by increasing imports from Saudi Arabia and Iraq, and later from Latin America (box figure), which (Venezuela's aside) were more expensive.[50]

Box figure 1.

India's major crude suppliers

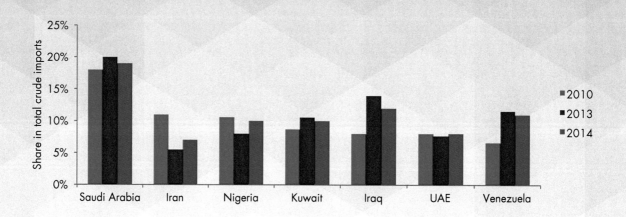

Payment problems also emerged after sanctions were imposed on the financial sector, after which the Reserve Bank of India scrapped the Asian Clearing Union payment mechanism. Thereafter, around half the payments (about 55%) were made through Turkey's Halkbank, and the rest were paid in Indian rupees through UCO bank, which could risk transactions with Iran due to its minimal exposure to international markets.[51]

Maritime (re)insurance was equally fraught. The ban on EU insurers and reinsurers from covering Iranian oil shipments, combined with Indian insurers denying or offering inadequate coverage, forced Indian refiners to significantly cut down crude imports from Iran.[52] Only a few Indian companies engaged with Iranian insurers, doubtful of receiving payments against any claims due to sanctions on Iran's financial sector.[53, 54]

The sanctions also halted Indian plans for an Iran–Pakistan–India natural gas pipeline as well as other initiatives of Indian companies (public and private) in exploration, production and infrastructure.

If the latest (mid-2015) deal between Iran and the P5+1 group of economies holds, sanctions could be eased, and India could start trading normally again with Iran. But the episode drove home India's deep supply vulnerabilities through its exposure to imported, particularly West Asian, crude oil.

India's energy demand is no longer marginal in global energy markets. It is the world's fifth largest producer of electricity, the second largest importer of coal, the world's fourth largest consumer of oil (3.73 million barrels a day)[55] and the 11th largest natural gas consumer with potentially much greater demand in future (it currently imports one-third of its natural gas consumption). More significant, as the world's fastest growing major economy, its primary energy demand will at least double by 2030 from 2011.[56]

China and India are leading the growth in new demand for oil and gas and other energy and nonfuel mineral resources. In 2009 China became the world's largest energy consumer. In the first decade of the 2000s the share of these two countries in global fossil-fuel trade more than doubled in value terms (to 10.8%) and tripled in weight terms (to 14.3%). Between now and 2030, however, Indian demand is projected to increase faster than that of any other country in the G-20. With China it will account for 51.6% (16.6% alone) of incremental global energy demand by 2035.[57] India, in other words, will get more deeply integrated into the global energy system.

Climate change complicates India's energy security. In 2010 it was the second most vulnerable in the world on an index of vulnerability (Bangladesh was first).[58] More recent studies find severe potential impacts of climate change on heat stress and human health (raising demand for air conditioning), food stress, water stress, river flooding, coastal flooding and infrastructure risks.[59] The unimpressive ambitions of China, the European Union and the United States to limit their greenhouse gas emissions have constrained the available carbon space for countries like

India. It is compelled to make choices about its energy investments after recognizing the consequences of global average temperature rise crossing the threshold of 2 °C, the international consensus among negotiating countries in the global climate regime.[60]

Energy security for India, then, is broader than merely reliable access to resources at a reasonable cost. Instead, a more appropriate definition would be the *availability* of adequate quantities of critical resources, at *prices* that are affordable and predictable, with minimum risk of *supply disruptions*, to ensure *sustainability* for the environment and future generations. But again, energy security is not the same as energy independence. India will not be energy secure by pursuing an autarkic policy. Instead, it has to build the infrastructure, financial, diplomatic, military and technical capacity required as its domestic energy system interacts more closely—in both directions—with global energy markets.

Seeking assured supply

In the first of the four imperatives, one approach that India has followed with some vigour is in trying to acquire acreages of oil and gas or ownership of coal mines abroad, driven by commercial and strategic considerations. Commercially, Indian companies have an interest in acquiring assets to widen their portfolio of reserves, forming joint ventures with foreign partners and getting access to technology. Such acquisitions would be evaluated based on the returns on investment. Strategically, the expectation is that owning energy assets overseas would buffer Indian consumers from supply and price shocks.

How have Indian companies fared in overseas asset acquisition? As the other major new energy demander, what has China's approach been and would it be applicable to India? To what extent has the strategy of Indian firms been effective? And does India (public and private firms and government) have the requisite capacity to secure overseas assets?

Public, private or both?

Oil and gas

State-owned enterprises have for the most part taken the lead in the oil and gas sector. Until 2013–2014, ONGC Videsh Ltd. (OVL)—the overseas arm of India's largest oil and gas company—made cumulative investments of more than $22 billion in 35 projects across 16 countries.[61] With this investment portfolio, proven reserves in OVL projects were 207 mtoe (as of 1 April 2014). In 2014–2015 OVL invested a further $4.7 billion in producing assets in Brazil and Mozambique, in addition to investing in four exploratory blocks in Bangladesh and Myanmar.

Other public sector enterprises—primarily Bharat Petroleum Corporation Ltd. (BPCL), Oil India Ltd (OIL), Indian Oil Corp. Ltd. (IOCL) and Gas India Ltd. (GAIL)—have made significant investments in upstream assets abroad. BPCL, through its wholly owned subsidiary BPRL, holds a participating interest in 11 exploration blocks in Australia, Brazil, East Timor, Indonesia and Mozambique.[62] It also plans to invest $2 billion in gas blocks in Mozambique and Brazil by 2018.[63] Similarly, OIL's overseas exploration and production (E&P) portfolio comprises 16 blocks spread over 10 countries (covering Bangladesh, Gabon, Libya, Mozambique, Myanmar, Nigeria, Russia, the United States, Venezuela and Yemen).[64] In one of the largest LNG projects, the Rovuma Area 1 field in Mozambique, state-run ONGC, OIL and BPCL have jointly invested more than $6 billion so far and plan to invest another $6 billion by 2019.[65]

Some Indian private companies, such as Reliance Industries and Essar Energy, have followed a similar trail. Reliance Industries (RIL), India's largest private sector firm by market capitalization, has so far invested $8.1 billion in its US shale gas business through three joint ventures, including $2 billion for stakes that give it access to 12 trillion cubic feet of shale reserves.[66, 67] Consolidating its overseas assets to boost its upstream asset portfolio, it has exploration rights for blocks in Myanmar, Peru and Yemen.[68]

Coal

Indian companies, mostly private, have purchased coal assets worth $12 billion over the past decade through nearly 20 deals in Australia, Indonesia, Mozambique and the United States (Table 5.2).[69]

In addition to investing in mergers and acquisitions, Indian firms such as Adani Enterprises, Lanco Infrastructure and publicly owned Coal India Ltd. had planned further investments (up to $32 billion) for mines and infrastructure development.[70] But after staring at huge losses on their overseas investments (due to low prices or poor quality coal),[71] many have been planning to sell off part of their stakes.[72] Coal India Ltd. had plans to invest $6.5 billion during the 12th Five-Year Plan (2012–2017) to develop Mozambican blocks and acquire new assets. Instead, it has decided to lower its risk exposure and allocate funds strengthening the domestic infrastructure for coal excavation and increasing supply.[73]

Table 5.2.

Major investments in overseas coal mines by Indian companies

Country and Indian enterprise	Investment
Australia	
Adani Group	$2 billion for Gailee coal block[a]
Lanco Infratech	$0.75 billion in western Australia
GVK Power and Infrastructure	$1.26 billion for Hancock coal[a]
Mozambique	
Coal India Ltd.	$180 million in A1 and A2 blocks
International Coal Ventures Private Ltd. (ICVL)	$50 million for 65% stake in Benga mines
Tata Power	35% stake in Benga mines
Jindal Steel Works (JSW)	$180 million for 90% stake in Tete province
Indonesia	
Tata Power	$1.3 billion for 26–30% stake in three coal mines
Adani Group	Bunyu mines
GMR Infrastructure	$550 million for 30% stake in Golden Energy mines
Reliance Power	$500 million in 3 coal mines
Essar Group	$118 million in Aries coal mines
United States	
Essar Group	$600 million in West Virginia

a. Cash and royalty deal for coal mines, port and railway infrastructure.
Source: Authors' analysis.

Lagging far behind China

India's efforts at overseas acquisitions are far surpassed by China, whose oil companies have been strategically diversifying their overseas portfolio and now have producing assets across 42 countries.[74] In 2010, Chinese overseas oil production was concentrated mostly in Angola, Kazakhstan, Sudan and Venezuela.[75] Since then, China has increased its footprint in Iraq and the Americas, to mitigate the risks in Sudan and South Sudan (Figure 5.1).

Figure 5.1

Chinese overseas oil production by region, 2013

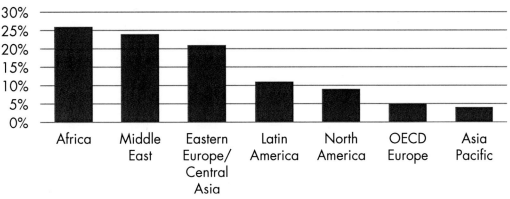

Chinese overseas oil production by region (2013)

■ Share of overall overseas production

Source: IEA 2014.

The scale of Chinese investments has been of a different order than that of Indian companies. Between 2009 and 2013 Chinese companies invested about $120 billion in global upstream merger and acquisition deals, five times the cumulative investments by OVL, which holds the majority of Indian overseas upstream assets in oil and gas (Table 5.3).

Table 5.3.

Chinese overseas investments in upstream oil and gas assets

Year	Investments in global upstream O&G assets (USD billion)	Overseas production by Chinese companies mboe/day (mtoe/year)
2009	18.2	1.1 (54)
2010	29.4	1.36 (67)
2011	20	NA
2012	15	1.8 (90)
2013	38	2.5 (123)

Source: (IEA 2011),[76] (IEA 2014).[77]
http://www.globalsecurity.org/military/world/china/oil.htm
Andrei V. Belyi, Kim Talus (2015) "States and Markets in Hydrocarbon Sectors," International Political Economy Series. Available at: https://books.google.co.in/books?isbn=1137434074

Besides the M&A deals in upstream assets, China has also heavily invested in loan-for-oil and loan-for-gas deals to secure long-term supplies. In 2009 and 2010, its oil companies signed 12 such deals in nine countries across Africa, Central Asia and Latin America for $77 billion worth of long-term loans (IEA 2011). It also invested $29 billion between 2011 and 2013 in loans for oil or gas deals with Russia and Turkmenistan (IEA 2014). China's government was active in securing these deals, while its national banks (China Export Import Bank and China Development Bank) supported the investments through long-term concessional finance.

As a third route for securing energy supplies, China's oil companies have signed multiple long-term liquefied natural gas (LNG) supply contracts with Australia, Iran, Malaysia and Qatar. Together, these contracts will ensure LNG supplies of about 43 BCM a year to China until 2032–2034 (IEA 2014).

State-owned enterprises dominate China's oil and gas sector, along with its publicly listed arms.[78] These companies account for almost all of China's overseas production (Figure 5.2). Since 2008, even smaller private companies have been investing abroad, though for far smaller stakes. For instance, five private firms invested $1.5 billion in 2013 to acquire upstream assets in Argentina, Canada, Kazakhstan and the United States (IEA 2014).

Figure 5.2.

State-owned entities account for most of China's overseas production

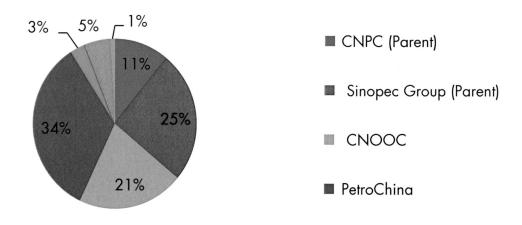

Source: IEA 2014.

Interest in overseas sources of coal has also rapidly grown in China. With coal consumption of 1,925 mtoe (in 2013), China accounts for half of global coal demand. It quickly became the largest coal importer after 2011 (from being a net exporter until 2009), despite also being the largest producer. The reasons? Higher costs of domestic coal, blockages in domestic transport infrastructure and environmental policy. By 2013 imports accounted for 8% of China's coal consumption; Australia and Indonesia were its chief suppliers.

Chinese companies have been more measured in overseas coal-mining investments than in oil and gas, though until 2010 large Chinese companies invested in coal mining in Australia and Indonesia. Once earlier plays appeared costly, the investments declined. And once the price of coal fell in 2013, overseas assets again piqued the interest of Chinese mining companies. Since then, Chinese firms have invested more than $15 billion in large mining and infrastructure projects, in Australia, Indonesia and the United States. Senhua, China's leading state-owned mining company, has signed major coal supply contracts with its neighbours, Mongolia (1 billion tonnes over the next 20 years) and Russia.

In summary, China's aggressive investments in overseas oil, gas and coal assets have been driven by the large financial reserves of national oil companies, access to concessional finance and diplomatic support from the Chinese government. But mere access to financial and diplomatic resources cannot justify acquiring overseas assets, so China, cognizant of commercial interests and market realities, is also developing its domestic energy resources.

An effective strategy?

Has the strategy to acquire overseas assets worked for Indian firms on the whole? Not really, when seen from various aspects. The first relates to overall production. As a share of overall oil and gas demand, India's overseas production was less than 4% in 2014–2015. In 2014–2015 OVL (owning a majority of India's overseas oil and gas assets) reported production of roughly 9 mtoe of oil and oil-equivalent gas.[79] If OVL's ambitious plans of raising the output to 60 mtoe by 2030 were to materialize,[80] India's overseas production as a share of overall oil and gas demand would likely triple to ~11.5% (excluding production from private owned assets).[81] Against these targets China's overseas production already stood at 123 mtoe in 2013 (Table 5.3) equivalent to ~18% of its net oil and gas demand.[82] China significantly outpaces India in acquiring overseas assets and securing long-term supplies, despite the fact that it has a higher share of domestic production in its overall oil consumption than India.

Second, Indian companies have had to compete with the much deeper pockets of Chinese state-owned oil companies, which often have access to concessional finance though Chinese banks. Over the past decade, India has lost multiple bids for overseas assets (~USD 17.5 billion) to China across Africa, Asia and Latin America.[83] Over the past decade the two have competed for oil and gas assets in Sudan (2002), Iran (2004), Angola (2004), Kazakhstan (2005), Ecuador (2005), Myanmar (2005), Nigeria (2006) and Russia (2006, 2008).[84]

From time to time, the two countries have tried to cooperate as well.[85] In 2004 they bought shares in Sudan's "Greater Nile Project." In April 2005 they issued a declaration that included cooperating on energy security, exploration and exploitation of resources in other countries. In December 2005 ONGC and the China National Petroleum Corporation (CNPC) joined hands for the first time to buy oil assets in Syria. In January 2006 the countries signed a Memorandum for Enhancing Cooperation in the Field of Oil and Natural Gas, which included upstream exploration and production, refining and marketing of petroleum products and petrochemicals, research and development, conservation and the promotion of environment-friendly fuels. The agreement also called for trading in oil and joint bidding in third countries.[86] Again in June 2012 ONGC and CNPC agreed to jointly explore assets in other countries and strengthen partnerships in Myanmar, Sudan and Syria. India's hope has been that it is "better to cooperate than compete."[87] But cooperation has for the most part been limited.

Third, India has to operate in politically fragile states and regions such as Iran, Iraq, Kazakhstan, Libya, Nigeria, South Sudan and Venezuela.[88] This is a challenge for China as well, but the tensions are greater when interests overlap and conflict. Joint investments with Vietnam for oil exploration in the South China Sea have created tensions with China.[89] So the government exercises tight control over overseas investments of Indian public sector oil and gas companies. To the contrary, OVL has requested that its board be given more financial autonomy to make decisions up to $1 billion (currently its powers are capped at $50 million).[90] However, Indian firms will need such financial powers backed up by diplomatic and military resources to reduce their risk exposure in less stable areas. In Kazakhstan and Venezuela, China has used the loan-for-oil and loan-for-gas routes for securing long-term supplies. It is unclear whether similar arrangements would be the best use of India's resources.

Fourth, little "equity oil" actually makes its way back to India but is instead sold on global markets. If prices in India are not in line with international prices, there are two consequences: selling in this way or at a commercial loss. Ownership of assets is thus unlikely to affect prices at the margin.

In sum, equity oil and gas investments have limited value but cannot be entirely dismissed if assured supply is a primary concern. In an emergency, however, energy security would be better assured by the ability to produce and deliver the energy resources.

Seeking safe passage

More than overseas acreages, a more crucial aspect of energy security for India will be to avoid disruptions in transporting energy resources. It matters little who owns the oil fields, gas wells and coal mines if resources are held up by poor transport infrastructure or geopolitical tensions. Transport blockages could occur on home soil as well, as the chapter on infrastructure highlights. Here we pay attention to two key vulnerabilities, which largely determine whether energy resources headed for Indian shores receive safe passage: control of or access to the shipping fleet, and security threats in the Indian Ocean and South China Sea.

Ownership of tankers

Given India's energy trade and its global share, the Indian shipping fleet is extremely small. In 2014 India had a fleet strength of 111 oil tankers (including crude oil and product carriers)[91] with a mere 9 million dead weight tonnage (DWT) for overseas trade (Table 5.4).[92] Contrast this with global oil tanker capacity of 491 million DWT (in January 2013). Even though India's share in global crude imports rose to 11% in 2014, it accounted for less than 1.8% of worldwide oil tanker capacity. Indian tankers carried only 16% of the total crude imported in 2012. This share has plummeted further as India's fleet strength and capacity of oil tankers has stagnated since 2012 while crude imports have been rising steadily.

China, however, has built substantial shipping capacity. Against a 12.6% share in global crude imports in 2012, China held 7.7% of the world's oil tanker capacity (~38 million DWT), and around 50–60% of its oil imports arrive on domestic carriers.[93] The shares are 80–90% for Japan and 70% for the United States (Table 5.4). Moreover, Chinese oil firms have focused attention on owning very large crude carriers (VLCCs). Already owning 70 of the global fleet of about 640 VLCCs, they have placed orders for 27 new tankers, nearly a third of the total outstanding orders for crude carriers in the world. China is also keen on building supertankers at home.[94]

The rapid rise in China's oil tanker fleet, particularly VLCCs, has been driven by two factors. One is to boost its shipbuilding industry, which China considers as a strategic sector.[95] The second is to use large tankers as offshore strategic storage facilities, which could be used to procure and store crude when it is cheaply available, and which could be mobilized quickly.[96]

Table 5.4.

Major crude importers and reliance on the domestic shipping fleet

Importer	Share in global imports[97] (2014)	Share of imports on domestic fleet (year in brackets)
United States	21%	70% (2012)
China	17.5%	50–60% (2014)
India	11%	<16% (2014)
Japan	9.6%	80–90% (2014)

Source: http://www.wantchinatimes.com/news-subclass-cnt.aspx?id=20121024000045&cid=1102

Building a domestic fleet for energy supplies has its own limitations. The expectation that a tanker fleet carrying domestic flags could deter a future adversary from interdicting tankers would hold only if the concerned country had the naval capacity to escort its fleet in a crisis.[98] The presence of a robust shipping industry (rather than energy security) or the strategic decision to boost the domestic shipping industry has historically governed the ownership of tanker fleets for overseas energy trade.

For natural gas India meets one-third of its demand through imports, all through the LNG route. India started to import LNG only in 2004. The Shipping Corporation of India is the only Indian company that owns LNG transportation capacity, though in a consortium with Japanese companies.[99] Of late, Indian shipyards have started to tie up with LNG carrier construction companies in Japan and South Korea to develop domestic LNG shipbuilding capacity.[100] But a domestic LNG fleet is not seen as a necessity for energy security. In general, companies importing natural gas prefer to charter sophisticated LNG carriers, which are fuel-efficient and economical. LNG shipping interests across the globe are mostly commercial and often tied to liquefaction projects.

A major determinant and potential limiting factor in India's imports of natural gas could be its LNG regasification capacity. Japan, the largest importer of LNG, also has the largest regasification capacity. China, the third largest and the fastest growing LNG importer, has been rapidly expanding its import capacity and currently has 20% of global LNG regasification capacity under construction.[101] India expects to quintuple its natural gas imports by 2030, from 13 MMT in 2013–2014 to 68 MMT.[102] Against this, India's annual regasification capacity is around 25 MMT (only 17 MMT operational) and is expected to increase to 55 MMT by 2025.[103] Floating storage and regasification units, which are cheaper and have lower gestation time than land-based plants, have been developed by a third of all LNG importing countries, globally.[104] India must also consider this technology while expanding its LNG regasification capacity. In fact, five such units (in aggregate more than 21 MMT a year) are being planned in India.[105]

Imports by pipeline are another route for natural gas. For instance, 50% of China's natural gas imports are through pipeline networks. India has also been negotiating or planning for pipelines from Iran (IPI, via Pakistan), Kazakhstan, Myanmar, Russia and Turkmenistan (TAPI, via Afghanistan and Pakistan).[106] A subsea gas pipeline from Iran and Oman to India has also been under consideration.[107] But geopolitics, security, pricing and operational issues have hindered rollout of any of these projects.

The contrast in ownership of shipping assets between India and other major energy consuming countries should not be read in an alarmist fashion. After all, India is not as large an energy consumer as China or the United States. The lesson, instead, is that as India's share in global energy trade increases, its demand for crude and LNG tankers will grow. If other countries are an indicator, Indian firms will over time need access to a fleet size sufficient to cover at least 50% of imported energy resources. This will require coordination between ministries and agencies responsible for oil, gas, coal, shipping, ports and commerce and industry. For India, shipbuilding capacity would also have to increase, as would access to finance to place large orders for the fleet.

Cooperating over security threats?

The intersection between maritime and energy security is a potentially serious source of friction. Of India's trade by volume, 95% depends on maritime routes. India's broader region encompasses three of the most important choke points for world oil flows: the Straits of Malacca, Hormuz and Bab el-Mandeb (between Yemen and Somalia). Seventy percent of India's oil imports come from the sea route in the Indian Ocean. Its coal imports also traverse sea lanes in the region. A growing share of energy will come from the east, including through the Straits of Malacca, which are also crucial for China's crude oil imports. Anyone controlling the Straits of Malacca and the Indian Ocean can disrupt energy transit for the growing centres of global energy demand.

India considers that it is a net security provider in the Indian Ocean, being the largest maritime power in the region. With a vast coastline as well as island territories in the Arabian Sea and the Bay of Bengal, India can monitor energy and trade routes better than almost anyone else. It has used this geographical advantage to expand its maritime partnerships with other countries in the region.

From convening the first Indian Ocean naval symposium in 2008 to rolling out an India-led maritime surveillance project (with coastal surveillance radar in India, Maldives, Mauritius, Seychelles and Sri Lanka) in 2015, India has actively led regional cooperation on maritime security in the Indian Ocean.[108] It has also been collaborating with other major naval powers for joint patrolling and military exercises on bilateral, trilateral and multilateral bases. For nearly two decades India and the United States have conducted regular Malabar naval exercises in the western Pacific, which included the Japanese navies for the first time in 2007, in recognition of the importance of securing energy supply routes.[109] India has also conducted joint naval exercises with the Republic of Korea,[110] France and Russia[111] and has trained Vietnamese forces in underwater warfare.[112]

The strategic relationship with other navies will be tested more in the South China Sea, which has large proven and probable reserves of oil and gas (Figure 5.3). It has also become a major source of geopolitical tension. The Spratly and Paracel Islands are the most disputed island groups in the sea, reflecting historical disputes and contested maps, with tensions more recently ratcheted up by the prospect of controlling large energy resources. Indian joint ventures with Vietnam for oil exploration would be at risk if such tensions boil over into conflict.

Another security concern for India would be uninterrupted access to the Straits of Malacca, through which 55% of Indian trade transits.[113] Even China imports more than 80% of its fossil fuel imports through the straits. To hedge its reliance on the straits, China has become more assertive in the South China Sea has raised its naval presence in the Indian Ocean. In addition, China has been building its naval capacity and is expected to have 78 submarines (12 nuclear), 80 medium and heavily amphibious lift ships and 94 guided missile boats.[114] Securing safe passage through the straits for continuing access to the South China Sea has thus become a major concern for India.

Figure 5.3.

Hydrocarbon reserves in the South China Sea

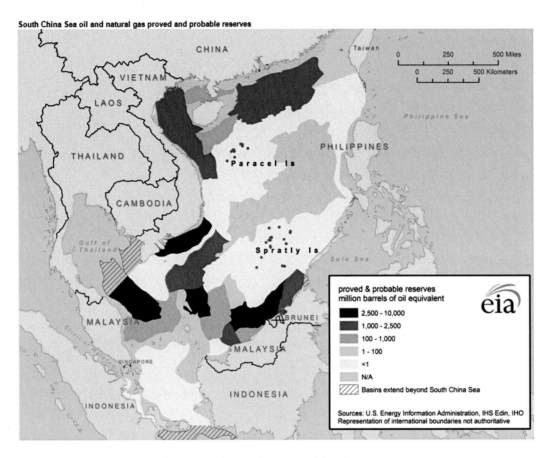

South China Sea oil and natural gas proved and probable reserves

Source: http://www.eia.gov/todayinenergy/images/2013.04.03/maplarge.png
This map is not to scale and does not depict authentic boundaries.

Beyond reasserting its position in the Indian Ocean, is India ready to assume a bigger security role in the greater Indo-Pacific region? In principle yes, but in practical terms it has some way to go.

India has been modernizing its navy. It has two conventional aircraft carriers, one amphibious transport dock, 50 frigates, destroyers and corvettes, one nuclear-powered and 14 conventional submarines, and many other naval and patrol ships.[115] To amplify its nautical capacity and reach, it expects delivery of seven stealth frigates, six diesel submarines and 30 other warships, as well as more than 150 fighters, maritime-patrol aircraft and helicopters over the next decade.[116] Over the last two decades, India has established a base for the Eastern Fleet (south of Vishakhapatnam) and upgraded its naval and air force facilities in the Andaman and Nicobar Islands.[117] It has also been strengthening its maritime and defence ties with its East and South Asian neighbours. It conducts regular naval visits and bilateral and multilateral exercises with regional states, such as Australia, Indonesia, Japan, Singapore, Thailand and Vietnam. And it has been playing a leading role in dealing with instances of disaster relief, piracy, smuggling, refugees and terrorism, thus presenting itself as a responsible and useful partner to South East Asian countries.[118]

But India or China cannot protect sea lanes unilaterally. First, both have traditionally been land-based powers, and it will take decades before their growing naval assets can underwrite the capability of maintaining a "blue water" navy.[119] Second, oil transit volumes through the Straits of Malacca alone will rise to 45% of global trade in 2035. Both China and India will have a strategic interest in maintaining the security of sea lanes. Rather than let their respective naval buildups become a source solely of suspicion, it will be important for the two navies to engage and interact more frequently. So even though strategic tensions are likely to remain between the two fastest-growing emerging economies, there is certainly scope, at a tactical level, to explore how joint exercises could help China and India maintain peace and stability over a region of mutual vulnerability.

Seeking secure storage

If avoiding disruptions in transporting energy resources is one part of securing supply, storing those resources is the other part.[120] Without adequate storage, sudden interruptions could have severe impacts on the domestic economy. The government knows this, but progress is painfully slow: strategic petroleum reserves were first mooted in India in 1998, but only in 2015 have the first storage facilities neared completion.

Answers to three questions could guide India in developing storage capacity and management practices. Is India's storage capacity adequate? Does it need to build more storage, or are there other options? And what institutional capacity does it need to manage the reserves?

Adequate storage?

The government set up Indian Strategic Petroleum Reserves Limited (ISPRL) in 2003 as a corporation to control and manage strategic crude oil inventories and coordinate the release of strategic crude oil stocks during supply disruptions, and their replenishment.[121] Strategic petroleum reserves (SPRs) capacity of 5.33 MMT (about 40 million barrels) has been developed at three locations: Visakhapatnam in Andhra Pradesh, and Mangaluru and Padur in Karnataka.[122] Taking advantage of low crude prices, India made its first purchase for SPR purposes in March 2015.[123] This capacity is expected to be commissioned and filled by the end of 2016. Additional SPR capacity of 12.5 MMT (93 million barrels), expected to be commissioned by 2020, is envisaged at four other sites in Gujarat, Karnataka, Odisha and Rajasthan. The storage at these sites would be in addition to the crude oil and petroleum products stored by the oil companies, which are not classified as strategic.

The current capacity of 5.33 MMT would provide at most 13 days of import protection at net imports of 148 MMT (2014–2015 levels).[124] With additional capacity of 12.5 MMT, this number would go up to 33 days (assuming that net imports grow at 5% on a compounded annual basis). By contrast, the United States has an SPR capacity of 97.7 MMT (727 million barrels), for import protection of 144 days at net imports of 5.04 million barrels a day (2014 levels).[125] China, by contrast, has strategic stocks of about 9 days worth of crude imports, with plans to raise its overall strategic capacity to 67.2 MMT (500 million barrels) by 2020. This would be equivalent to around 60 days of net imports, which may climb to 8.4 million barrels a day by 2020.[126]

India has storage facility in tanks and pipelines for not more than 45–50 days (Table 5.5).[127, 128] By comparison, IEA members have a minimum stockholding obligation (including strategic, commercial and operational stocks) of 90 days (of daily net imports) of all oil.[129] India, too, wants 90 days of stocks, but deployment over the past decade has been very slow. For the medium term at least, India's current storage capacity is expected to fall well short of IEA standards.

Table 5.5.

Storage capacity of top energy consumers

Country	Closing oil stock (in days of net imports)(December 2014)	Rank as primary energy consumer (2014)
China	22.7	1
United States	251	2
India	45	4
Japan	157	5
Canada	Net exporter	6
Germany	140	7
Korea, Rep. of	233	9

a. Including strategic and commercial reserves.
Source: IEA (2014) "Closing Oil Stock Levels in Days of Net Imports," December, available at http://www.iea.org/netimports/?y=2014&m=12; BP (2015) "Statistical Review of World Energy," June, available at: www.bp.com/content/dam/bp/pdf/Energy-economics/statistical-review-2015/bp-statistical-review-of-world-energy-2015-full-report.pdf

Is additional storage the only way?

The inadequacy of storage capacity, whether SPRs or other reserves, does not mean that India has to build all its required storage. Any new construction entails heavy financial commitments: every additional 5 MMT capacity (~13 days of storage) could cost Rs. 4000 crore ($615 million) in infrastructure costs alone.[130] Crude for filling these SPRs would cost another $2 billion for every 5 MMT (assuming a crude price of $50 a barrel). IEA member countries have most commonly used government budget support, bank debt and bonds to build SPR sites. For operation and maintenance most countries' SPR facilities rely on budget support but also impose levies on industry (which also contributes to compulsory stocks as well as to commercial stocks) and taxes on consumers. The large sums mean that India would need to think carefully before investing in infrastructure on this scale.

An alternative route to building storage facilities is to site some emergency crude oil stocks in other countries. For OECD countries this is a normal practice. The United Kingdom holds 31% of stocks abroad, and Belgium, Estonia and Ireland each hold more than a third of their stocks abroad (Table 5.6). Such storage is usually split among three or four countries to reduce risks of non-recovery in times of crisis. Once the share and location of overseas storage are decided, bilateral agreements have to be signed to formalize terms of storage and release. India could explore similar arrangements with its major oil suppliers. Despite the political risks, this is one way to quickly build up a strategic reserve and reduce vulnerability to supply disruptions.

Another possibility is to rely more on market mechanisms by signing long-term commercial contracts as forward options between exporting countries and refiners in India.[131] In signing such contracts it would be useful to seek flexibility in the form of fixed and optional volumes, which would help India to tap spot crude in the future whenever economical, the case today.[132] India has already invited oil-producing West Asian nations such as Saudi Arabia and Oman to invest in its SPRs.[133] Such a mutually beneficial strategy would augment India's storage capacity.[134] The arrangements would give major exporting countries a stake in India's stockpiling programme. Additional agreements could enable oil-sharing during emergencies or swap arrangements with countries even farther away.

What institutional best practices?

India could also learn from other countries' best practices for managing SPRs and emergency crude oil and petroleum products reserves. The US Department of Energy's Office of Petroleum Reserves, under the Office of Fossil Energy, manages three emergency stockpiles including the SPRs. With hundreds of staff and contractors, it offers services to support the SPR programme and involves private firms in operating and maintaining the reserves.[135]

Table 5.6.

OECD countries with emergency oil stocks held abroad (December 2014)

IEA country	Closing oil stock (days of net imports)	Stocks held abroad (industry and public)	Share of stocks held abroad
Estonia	346	138	40%
Luxembourg	92	76	83%
Belgium	163	65	40%
United Kingdom	196	60	31%
Ireland	114	39	34%
The Netherlands	167	36	22%
Italy	123	20	16%
New Zealand	97	17	18%
Austria	110	13	12%
Sweden	134	12	9%
Czech Republic	139	11	8%
Switzerland	159	11	7%
Portugal	103	7	7%
Germany	140	6	4%
Finland	249	4	2%
France	111	3	3%
Spain	116	2	2%

Source: IEA (2014), Closing Oil Stock Levels in Days of Net Imports, June (Available at: http://www.iea.org/netimports/?y=2014&m=12).

In China the National Oil Reserve Centre is the core body responsible for SPR construction and oil procurement. The National Oil Reserve Office of the National Energy Administration (NEA), in turn, oversees the Centre. At the apex is the State Council, which can order releases from the SPRs, implemented through coordinated action between the NEA, the Ministry of Finance and the National Development and Reform Commission. Major national oil companies, such as CNPC, Sinopec and Sinochem, serve as contractors for operations of strategic oil reserves. While building the SPRs the Chinese government is encouraging state-owned oil companies to increase their commercial reserves. It is also considering placing a minimum stockholding obligation on industry, thus creating its National Petroleum Reserve, which will include crude oil and petroleum products.[136]

India has drawn on the OECD's practices of establishing oil emergency response organizations, stockholding oil and implementing oil-stock drawdowns and other emergency response measures. But the authorities, mandates and coordinating mechanisms to respond to energy-related emergency situations require more clarity.

The National Crisis Management Committee is the topmost executive committee for coordinating responses to all kinds of emergencies.[137] For stockholding oil ISPRL is India's nodal agency. The Petroleum and Natural Gas Regulatory Board (PNGRB), an autonomous body, has the mandate of ensuring uninterrupted and adequate supplies of petroleum, petroleum products and natural gas in all parts of the country. During an emergency, the government can invoke Section 43 (1) of the PNGRB Act 2006, which allows it to take over control of the entire downstream sector. But it is unclear how the institutions mandated with crisis management would coordinate with ISPRL in any event of oil supply disruption. There is a need to clearly outline an emergency response mechanism that India would follow. How would coordination between the government, the oil marketing companies and ISPRL be accomplished? And how would SPR levels be replenished once normalcy is restored?

In addition to the release of oil stocks, other emergency response measures include demand restraint and fuel switching.[138] The government can invoke the Essential Commodities Act to maintain the equitable distribution of petroleum products. But there is no demand management plan that could be put into use during any supply disruption. To this end India needs a contingency plan to restrict demand during an emergency and encourage fuel switching. It could cover compulsory use of public transport or carpooling or curbs on using diesel-based generators.

Refined products should also be available in case of emergency. Public stocks held by government or a stockholding agency in IEA countries mostly consist of crude oil. But two-thirds of IEA countries impose minimum stockholding requirements on their industry,[139] indirectly ensuring that some refined products are available for emergency purposes. The approach also establishes an operational link between a country's reserves and its oil companies to help ensure rapid drawdowns of resources in an emergency.[140] India has no such requirement on its industry. To this end, India would need a specific law, such as an Oil Stockholding Act, which could make it obligatory for upstream companies to keep emergency crude oil stocks and downstream companies to keep emergency stocks of petroleum products.

SPRs no longer merely ensure that oil reaches refineries. Since the United States has become a swing producer in global oil markets, there are calls that it uses its SPRs to ensure that oil reaches refineries before a global oil price shock damages the United States and the global economy—by freeing up more oil for global markets.[141] India could also use its SPRs strategically to guard against severe price shocks.

Seeking international cooperation

Equity investments in energy resources overseas, boosting merchant shipping fleets and naval capabilities and building and managing strategic reserves—all are part of the apparatus that India will need as it integrates more deeply with the global energy system. But these measures have their limitations, exposing the extent to which Indian firms and the government can make the best use of opportunities in energy markets or withstand shocks. India is not unique in having these vulnerabilities. Energy consumers and producers worldwide have to become more resilient to fluctuations in global energy markets, because the upside of securing resources at the most optimal prices remains highly attractive. The role of international institutions is to help their member states achieve goals that countries would find difficult to achieve on their own. But what energy-related institutions could India work through to increase its resilience and take advantage of the benefits that deeper global integration provides?

Little capacity to be involved in multiple forums

India faces a tough choice when shaping its strategy on engagements with international institutions. Despite its growing economic and political power, its position in institutions will depend on how such engagement can improve the human condition of its people. This is the guiding principle of its diplomacy.[142] At the same time, India is not satisfied with the status quo at international institutions, which freezes its position as a lesser power and limits its room for manoeuvre. Yet when it makes demands for reform, it is often asked about the kinds of responsibilities it is willing to assume. This dichotomy means that India often lacks a clear strategy in its international engagements.

A related challenge is the limited capacity in the Indian system to deal with the growing complexity in energy and resource issues. Energy alone cuts across line ministries for oil and gas, coal, renewable energy and nuclear power. At home many of the energy choices would also depend on decisions by other ministries, such as finance and water, and by state governments. Along the land or sea border the role of the military as well as coastal and port authorities becomes crucial. Internationally, commercial and political diplomacy has to be carried out through bilateral channels and in multilateral forums. Global climate change and its regional and local impacts add further complexity. All these issues touch on other contentious international policy challenges such as trade, finance and technology transfer.

Despite these challenges, India has no option but to engage with energy-related institutions. As argued earlier it is no longer a marginal energy consumer that keeps it immune to global energy shocks. India's vulnerabilities are not macroeconomic alone—they are systemic. Equally, it does not have the base of domestic resources and diplomatic and economic weight at international level so that it can alone try to shape the direction of political and market outcomes. Given that its demands from the international energy system will only grow, India has a strategic interest in well-functioning energy markets across the world.

The trouble is that there is no overarching global regime for energy. Energy governance is highly fragmented, with neither India nor China as members of the IEA.[143] Yet there are multiple overlapping institutions relevant to global energy governance, including the IEA, the World Trade Organisation (WTO), the United Nations Framework Convention on Climate Change (UNFCCC), the Energy Charter Treaty and the Asia-Pacific Economic Cooperation (APEC) Energy Working Group, which comprises the region's major energy consumers (but not India). A world with multiple poles of energy suppliers, energy demanders and emerging economies has direct implications for coherence among the international bodies.[144] The countries belonging to the multilateral trade regime do not always overlap with those belonging to producer cartels. There are new calls for bringing together major suppliers and users under a global Energy Stability Board (similar to the Financial Stability Board) to coordinate emergency actions and give voice to emerging economies.[145] But it is unclear which forums countries will choose to resolve disputes.

Choosing functional institutions

India's participation in energy regimes is limited. It is not a member of most energy-specific regional organizations: it has only observer status at the IEA, since it is not a member of the OECD; and because it is not a member of APEC, it is not a member of that body's Energy Working Group. India will need allies in these forums. For instance, as a member of the G-20, which has rotational presidencies, India could use (at a future date) its presidency to introduce energy security into the agenda and shape international cooperation. How should India choose which forums to engage?

India has an interest in working with other second-tier energy demanders, especially among Asia-Pacific countries, either to shape existing institutions or create a regional energy regime. The choice of institution should not be dogmatic. It should be driven by the need for certain core institutional functions to be delivered efficiently and effectively.

The first task would be to increase transparency in energy markets with regular information on oil and gas purchases, long-term contracts and spot market prices. This function would help when Indian firms are trying to renegotiate contracts, make best judgments about oil futures and reduce the premiums they pay for crude shipped to Asia. The Joint Organization Data Initiative has since 2002 helped coordinate efforts among six international institutions to create awareness for greater transparency in oil markets—APEC, OPEC, the IEA, the Latin American Energy Organization, the Statistical Office of the European Communities (Eurostat) and the United Nations Statistics Division. This could serve as a model for drawing in new energy demanders in the Asia-Pacific region as well.

Second, membership in an energy institution should facilitate discussions on how each member country's strategic reserves could instil confidence in energy markets to mitigate short-term supply shocks. This is the premise for IEA countries to support each other during emergencies, and why many European countries have strategic reserves outside their territories. For India, mere membership in the IEA would not guarantee physical supplies of oil and gas during an emergency, given geographical limitations. But a regional platform could be more suited to that purpose, drawing on best practices adopted elsewhere.

Third, any energy-related institution that India joins ought to discuss the protection of key energy supply routes (by land and sea). Joint naval exercises are one way to provide security. But overall security will need much more cooperation among members, including intelligence sharing, real-time monitoring of oil and gas tankers and bulk carriers, and quick emergency-response measures. Such protocols could be established in the short term, and trust built along with standard operating procedures in the medium term.

Fourth, in a more institutionalized form, energy-related institutions should be able to arbitrate disputes and protect overseas investments. Existing institutions—such as the WTO and international arbitration courts—cover many of these aspects. But the rules are often conflicting, as with the way rules on climate change contradict how the WTO treats subsidies.[146] India has interests on both sides. It is keen to attract foreign investment to its domestic energy sector, but it must protect its investments in other countries, particularly where regional and geopolitical tensions are on the rise. Moreover, lack of clarity on trade rules continues to stymie cross-border investments in emerging energy sectors. India will have to choose the forums for dispute resolution carefully.

Fifth, India could explore how a group of countries could pool resources to acquire assets or to reduce the competitive premiums they might be paying individually to purchase overseas assets. These concerns might not seem large when the global energy markets are showing signs of oversupply. Yet collective bargaining would have merits when market conditions change, demand rises or new security threats emerge.

Sixth, India's membership in an existing or new energy institution should provide it with a platform to share experiences with other emerging economies and demonstrate leadership in areas where its strengths lie. These areas include the growing emphasis on renewable energy, the technical and financial innovations in decentralized energy, the R&D efforts in energy storage and the unique pressures of rapid urbanization on an unprecedented scale. India has committed to increase its installed capacity of renewable energy from about 35 gigawatts (GW) today to 175 GW by 2022.[147] It also has more than 400 firms operating in the decentralized energy market, delivering energy services to poor communities.[148] And it is keen on R&D in energy storage, as outlined in its recent submission of commitments to the UNFCCC.[149] Then there is its Smart Cities mission to upgrade the infrastructure in 100 cities across India, with significant emphasis on energy for buildings and transportation.[150]

Finally, India could consider new institutions where its interests may be limited currently, but that deal with issues having a direct impact on its strategic interests. One is the Arctic Circle Forum, which has already attracted China's attention. Oil and mineral exploration in the Arctic, a consequence of a warming climate, will also drive further greenhouse gas emissions. The opening of new shipping routes will create new trade opportunities as well as regional tensions. Access to these regions could become restricted to a limited set of countries, excluding others from both governance and the mineral resources. Of critical interest to India, such issues would demand regular engagement, at least as an observer.

Conclusions

India's integration into global energy markets will be one of the key shifts in the global economy in the first half of the 21st century. The routes India chooses will determine how energy (in its various forms) will drive its economic growth and the overall human development of its people. As a major energy consumer and a growing energy importer, India faces tough choices. Autarky is not an option, given the scale of demand and the rising aspirations for higher standards of living.

This chapter has explored four imperatives from the perspective of India's quest for energy sources in a manner that is affordable, has minimal supply disruptions and is consistent with the need for environmental sustainability. All four have limitations, partly due to capacity constraints at home and to India's still-limited power on the global stage. But in combination they would make Indian firms more adept at exploiting opportunities and its energy system more resilient to shocks.

Some unknowns remain but will continue to shape India's integration into the global energy system. First, what will be the medium-term price of oil? This will be the defining question for any Indian policymaker. Yes, there is oversupply today, but demand will rise as emerging and other developing countries pick up their pace of economic growth. At what level will the price of oil stabilize, and when, are questions that require continuous assessment.

Second, will there be a global price on carbon, and what will it be? The answer depends on many factors: global climate negotiations, regional carbon trading schemes, national policies and internal carbon prices, which many firms (including the largest energy companies) are already using. Without clarity on the price of carbon it will be difficult to make long-term investment choices or protect the value of existing assets. This uncertainty will affect not just energy reserves but also transport infrastructure, storage facilities and distribution channels.

The biggest unknown is the relative roles of the public and private sectors in India's growing energy system. Government policies and regulations will shape the investment climate at home. Public sector energy firms will compete, within their limited mandates, for overseas energy assets. Private firms with interests in fossil fuels might attempt the same, but would they compete or cooperate with state-owned enterprises? Will India's diplomatic corps and other state agencies support India's firms (public or private) in their dealings in overseas markets?

Private firms in the growing renewable energy sector will have their own business interests at home and abroad, relating to investments in technology, manufacturing facilities and services. How will India's policy apparatus manage conflicts between these different interest groups? Will an energy pricing regime in India give credible signals to investment needs, and to climate and local environmental sustainability? And will public and private firms (both Indian and foreign) operate in a regulatory environment where their investments are secure from arbitrary changes in policy? These questions and uncertainties go well beyond energy security. But their answers will determine the background for India to prepare itself for a leading role in global energy.

CHAPTER 6

Politics and Policies for a Resilient and Equitable Energy System

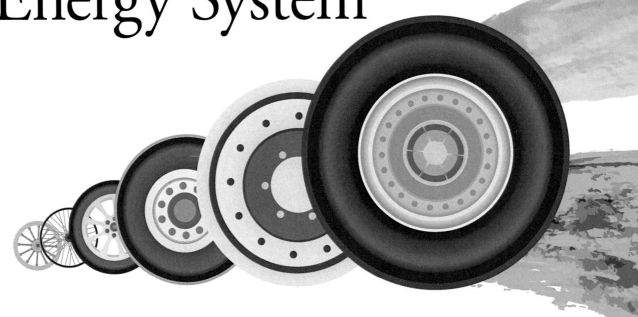

Inclusive and rapid economic development—a clear priority of the Indian government—brings with it the imminent need to raise sharply the share of clean and renewable energy for a sustainable energy future. India will have to balance pressure on energy needs from a fast-growing economy with a large underserved population, as it continues relying on fossil fuels for the next decades. But boosting the supply of energy in an equitable, affordable and sustainable manner is not the only goal.

The discussion in earlier chapters has shown that for a more energy-secure future, India needs to pursue several other goals simultaneously:

- Reducing the share of fossil fuels in the primary energy supply energy mix
- Increasing the share of renewables
- Reducing exposure to geopolitical risks by diversifying the fuel basket as well as import sources
- Reducing the energy intensity of economic output and increasing energy efficiency across all sectors
- Providing reliable access to electricity and modern cooking fuels and technology; and using available resources optimally

India's energy policy-making has multiple objectives: some are complementary, while some may introduce trade-offs. Economic development, urbanization, growth of the manufacturing sector, etc., would require the setting up of massive infrastructure and call upon technologies both on the supply and demand sides to minimize the implications of energy and infrastructure development on land, water and air. Moreover, policymakers would need to grapple with the trade-offs of balancing costs that the optimal solutions may entail with the available budgets.

So, in working towards a sustainable long-term energy vision, planners and decision makers have to resolve the trade-offs in an integrated policy and regulatory regime, providing the right physical and financial environment for investment in clean technology and manufacturing industry. Such a regime is crucial for providing a strong and consistent direction to producers and consumers of energy, facilitating investment in the desired choices and creating the right markets in a timely manner. It also requires comprehensive planning across economic sectors to ensure cohesion, considering all aspects of reliable and affordable energy access to all population groups, including the local and global environmental effects of energy use. Policies geared towards incentivizing particular options need to ensure that the overall policy environment for the option is consistently supportive.

India's current energy policy and planning landscape is very diverse, horizontally, across several ministries—and vertically, often divided at central and state levels. So although several elements require cross-cutting approaches across ministries and departments, decisions more often than not are made in silos. Similarly, policy linkages are multifaceted, and some decisions in the energy sector are strongly tied to trade and foreign policy, such as accessing overseas assets, securing supplies and tackling climate change—all requiring coordination.

Besides adopting a holistic policy approach that addresses the complex linkages and trade-offs in different sectors, the government needs to closely align short-term plans with longer term planning. One example is the link between energy security and emissions mitigation. Here, India's planning should go beyond a multidimensional focus to be resilient, anticipating energy and infrastructure demands and responding appropriately with a suite of policies that enable rapid and sustained economic growth.

TERI's 2014 analysis of energy scenarios reveals strong directional synergies between the actions to move along pathways focused on improving energy security and those geared to mitigate emissions. Mitigation options, such as efficiency measures on the energy demand or supply side, bring additional benefits of reducing overall energy requirements and thereby easing the pressure on fossil fuel imports. Similarly, decentralized renewables—apart from providing clean energy—provide energy access in remote areas, thus improving energy access and security.

Against this backdrop, the policy space needs to be aligned along several tracks. First, make available adequate and reliable supply of fuels to meet the increasing demands of a country moving ahead. Second, ensure that the supply of energy is affordable to all sections of society. Third, provide incentives to options that are more efficient and clean, giving clear signals of these incentives to consumers and to producers. And fourth, ensure that a robust regulatory system is in place to effect the workings of these policies. Institutionally, government departments and ministries, at central and state levels, will therefore also need to coordinate closely.

This chapter makes the case for stable and time-consistent policies, reflecting the running debate over the licensing regime for exploring hydrocarbons and recent scandals over awards of coal-mining blocks. It next argues for an integrated policy environment for managing the demand and supply sectors, including infrastructure, and for resolving overlapping institutional authorities. It then presses the case for an independent regulator as the country gradually liberalizes its energy market.

The need for stable and consistent policies

The licensing regime for exploring hydrocarbons

Blocks for oil and gas exploration are contracted out under the New Exploration Licensing Policy (NELP). Nine NELP rounds have been conducted so far, with 254 blocks given out to companies for E&P. Under the NELP regime, production-sharing contracts (PSCs) are signed between the government and oil companies, which split production after the company recovers its costs (of exploration, development, production and royalties). The split share of post-recovery costs is biddable by the oil company, and the winning bid is determined by a host of parameters, including the share of profits the company offers to the government (Box 6.1). Apart from accruing revenues through profit sharing under the PSC, the government has a system of royalties and taxes. For coal bed methane, production-linked payment systems are followed.

Globally, PSC regimes have become more popular since Indonesia introduced them in 1966. But most large producers of hydrocarbons,[151] including the OPEC countries, either do not employ PSCs or have revenue-sharing PSCs, which in India are referred to as revenue-sharing contracts (RSCs). In 2010 Brazil proposed a profit-based PSC and the first round of blocks went into bidding in 2013, although these profit-sharing contracts are yet to be tested there. Profit-sharing PSCs have emerged as the regime of choice in most developing countries, including Angola, Oman and Peru, apart from India. Around the world, more countries have adopted profit sharing rather than revenue sharing as the guiding PSC framework.

The PSC debate in India

In an attempt to increase government revenues, India's finance minister in 2013 announced that the country would move from a profit-sharing to a revenue-sharing regime—that is, from a PSC to an RSC system. Such contracts would see production split right from the moment the oil companies start earning revenues, even before costs are recovered. RSCs had received the endorsement of the Comptroller and Auditor General of India in 2012. Another high-profile endorsement came in 2013 from a committee headed by Dr C. Rangarajan, then the chairman of the Economic Advisory Council to the Prime Minister.

Box 6.1

Private participation is on the rise in oil and gas

In the oil and gas industry, dominated by state-run exploration and production (E&P) and refining companies, the share of private firms has increased over the past decade due to a policy that sought greater private participation and thus the entry of new players (box figures 1 and 2).

Box figure 1.

Domestic production of crude oil by company type

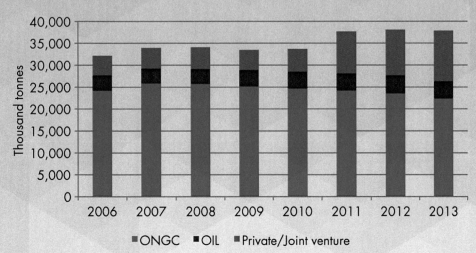

Source: MoPNG 2014.

Box figure 2.

Production of petroleum products by company type

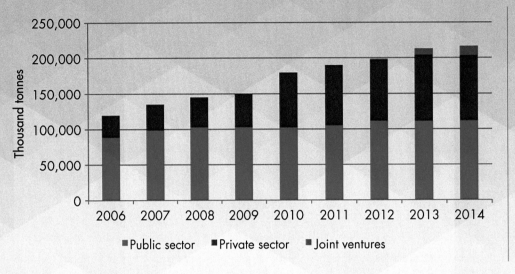

Source: Ministry of Petroleum and Natural Gas (MoPNG) 2013.

Due to differences in opinions between stakeholders on the recommendations of the Rangarajan committee, another committee was formed under the chairmanship of former Finance Secretary Vijay Kelkar. In 2014 the Kelkar committee, through a report "Roadmap for Reduction in Import Dependency in Hydrocarbon Sector by 2030," recommended that India continue the PSC regime. No change has in fact been brought in, and the debate is far from settled, as a new "draft revenue-sharing contract" may replace the NELP's PSC. The new contract purports to bring greater financial clarity for the government, as well as more revenue. An escrow account is sought to be created into which all oil and gas revenues would flow. RSCs draw support from the perspective that companies have no incentive to cut costs under the PSC. It is also felt that companies have an incentive to over-invoice on costs, denying the government some legitimate revenue. Under the revenue-sharing regime, by contrast, governments do not have to monitor activities, since the system encourages efficiency and cost-cutting.

Weighing in on the debate

Governments worldwide use a variety of fiscal regimes to safeguard oil and gas revenues. Instruments include concession agreements and joint ventures with national oil companies, as well as PSCs. Tax and royalty systems—also varying— often resemble PSCs. They are also driven by the state's desire to maximize revenues, but may have other outcomes on efficiency and may influence incentives in a different way. But these distortions are hard to quantify because nearly all countries have combinations of various taxes, royalties and production-sharing frameworks.

Three key claims are made in support of RSCs. First, companies do not produce optimally under the profit-sharing system (since they do not invest in cost-cutting mechanisms). Second, companies may cheat by overinvoicing under the PSC regime. And third, the government (with RSCs) would not have to monitor companies' production and expenditure as they attempt to safeguard revenues.

These points of contention in the debate emanate from the perceived incentives created by the two systems. But proponents of the PSC system claim that, owing to the very risky nature of petroleum E&P, RSCs do not provide incentives to invest. They also claim that fears of inefficiency and cheating are not grounded in reality, since the sector is very competitive and companies cannot risk inflating costs.

A lack of information on production potential, and on E&P cost projections before activities begin, makes either side's claims hard to pin down. It also magnifies the contractual risks. In India, the sedimentary basins are underexplored (Figure 6.1), and the domestic production of natural gas has been generally overestimated. Such overestimation can influence a government's decision on what the winning bids would be in a round of E&P block auctions. Further, nine of ten exploration efforts typically result in a loss in some regions around the world, though not specifically India.

Figure 6.1.

Oil and gas exploration performance in India's sedimentary basins

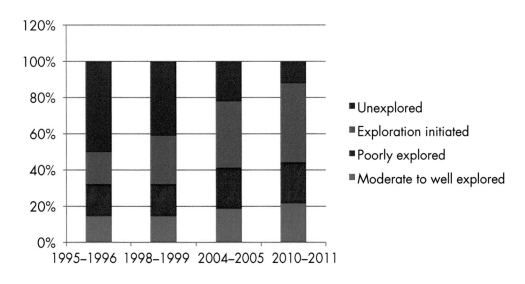

Source: Directorate General of Hydrocarbons 2014.

Experts also doubt the claim that the uncertainty of reserve estimation diminishes as the field reaches maturity and more data become available. But concern about companies underreporting reserve estimates is growing since it enhances their bargaining power at contract renegotiation.

In an RSC where the government's take is 60% and the costs turn out to be 40% of the gross revenues, the government's share would effectively be 100% of revenues after the E&P company recovers its costs. If the government's take is proportionately higher, the company would effectively take a loss, discouraging it from developing the field to its full potential.

Companies have incentives to cut costs under both systems, as cost overruns affect every operation, and no company would like to risk deliberate overspending or unexpected cost inflation. Further, overinvoicing is difficult to conceal, and could be extremely risky. Monitoring mechanisms—including third-party auditing, authorization for expenditures and other measures—can ensure that there is no cheating.

But there is a genuine risk of opportunistic gold-plating of costs when the revenue structure follows a step-scale—for instance, a 60% government share when the rate of return of company is 10%, an 80% government share when the rate is 20% and so on. Increasing costs or reducing production just before the step up to a higher government share can benefit companies at the expense of the government, though its impact can be mitigated with a smoother, formula-based approach to profit sharing.

Recent events—and the way forward

Governments have an incentive to secure and then boost revenues from the oil and gas sector. Most countries, including India, deploy a cocktail of measures to extract rents from the sector, including royalty payments, taxes and PSCs. There is an added dimension of import dependency in India. The Kelkar Committee Report points out that the average government take in exporting countries is much higher, at 71%, than in importing countries, at 52%.

Aiming to increase its revenues from the sector, the Indian government declared in September 2015 that it would be deploying RSCs for 69 marginal fields it intends to auction soon.[152] Evidence from these 69 blocks could throw light on interest from the industry, the need for a mechanism of checks and balances and other aspects of the revenue-sharing mechanism. It would also provide a template for the government to help it decide whether RSCs can be extended to all future block auctions of hydrocarbons. Profit-sharing and revenue-sharing regimes have advantages and drawbacks. Either way, whatever the evidence may show, the government must have a consistent policy for the next few decades, given its wider energy-security objectives, developing a robust mechanism of checks and balances to ensure transparency in production.

Rationing domestic national gas

Introduced in 2008 the Natural Gas Utilization Policy ranks gas-consuming sectors to ration domestic gas. The government felt this was necessary in light of growing domestic supply from the Krishna-Godavari-D6 fields,[153] even though demand remained greater than domestic supply. Contractors of blocks under the New Exploration Licensing Policy (NELP) are obliged to follow this policy when supplying products.[154] The order of priority is existing gas-based urea plants, existing gas-based LPG plants, gas-based power plants lying idle or underused, city gas distribution networks for piped natural gas to households and compressed natural gas to the transport sector, and finally gas-based power plants.

The government chose this order because the 22 natural gas-based urea plants could not meet demand and had to use naphtha and fuel oil, both costlier alternatives. Further, imports met 25% of India's LPG demand. And while the government recognized the importance of city gas distribution projects, petroleum and natural gas did not have high enough penetration in the country. The Natural Gas Utilization Policy was later modified to include non-priority sectors, including steel plants, refineries and petrochemical plants. After a fall in output from the KG-D6 block, the government reduced allocations of gas to noncore sectors.

The Natural Gas Utilization Policy—an outcome of the inadequate supply of domestic natural gas—has ended up creating distortions of its own. Priority sectors such as power and fertilizers have controlled consumer prices and depend on low-priced domestic natural gas. Sectors such as city gas distribution and industry, which can absorb higher prices, come lower down the priority list. The government is considering revising the policy in light of changing priorities, which include greater power supply and improved availability of petroleum and natural gas. It also, in May 2015, mandated pooling prices for LNG and domestic gas for use in the fertilizer sector, and it is trying to revive gas-based power generation by providing subsidies. While in the short run a subsidy is provided, the cost of power generation from underground domestic coal is likely to be higher than domestic natural gas and comparable to LNG. Accordingly, the government should consider pooled pricing for the power sector as well.

Although the sector contributes to the government's subsidy burden (Rs. 70,967 crores in 2014–2015), fertilizer companies have been importing natural gas to make urea. In April–December 2014, 26 million standard cubic metres per day (mscmd) of domestic natural gas and 16 mscmd of imported LNG were used. To lower the subsidy burden, which is higher for imported urea, the government took a step to increase domestic fertilizer production. In March 2015 the Cabinet Committee on Economic Affairs approved price pooling for domestically produced and imported natural gas, to provide domestic manufacturers with better price signals. Such price pooling of gas could save Rs. 1,550 crore in subsidies between 2015–2016 and 2018–2019.[155] Given the potential importance of natural gas in the energy sector, policy should be directed towards pooling gas prices to enhance the relative attractiveness of gas as one of the strands of moving to a more sustainable energy sector over the long term.

Coal mining

Production

India's energy sector is dominated by coal on the supply side. Like consumption, production has continuously increased, although the share of imports has also risen over the years (Figure 6.2). With coal demand expected to reach 1.2 billion tonnes in 2019–2020, the government has announced that it expects Coal India Ltd. (CIL)—the sector monopoly—to increase production from around 462 MT in 2013–2014 (CCO 2014) to 1 billion tonnes by 2020.

Figure 6.2.

Coal (coking and noncoking) production and consumption (million tonnes)

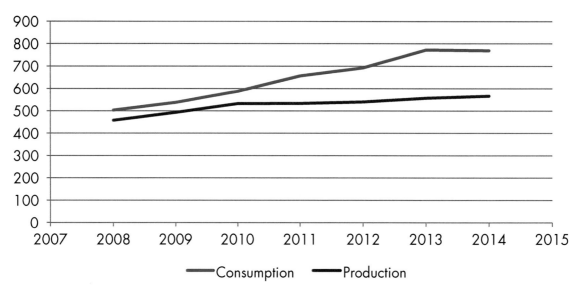

Source: Coal Controller's Organization 2014.

To meet this target, CIL is planning to purchase 2,000 rail wagons for evacuation and upgrade technology in opencast mines. The latter includes high-capacity equipment, an operator-independent Truck Dispatch System, a vehicle tracking system using GPS/GPRS, coal-handling plants and SILOS for faster loading and monitoring using laser scanners.[156] It aims to ramp up production from underground mines using newer technology such as continuous miner technology at large scale, long-wall technology in some places, man-riding systems in major mines and telemonitoring.

But it is expensive to upgrade underground technology, because India froze it in the 1970s, and upgrading would require imports. Hardly any new capacity in underground mining has been created in the last several decades: underground mines contributed a mere 8.8% of India's total raw coal production of 516 MT in 2013–2014, opencast mines, 91%.

The main reason was an implicit regulatory push towards opencast mining in the 1970s and 1980s, such that the share of coal production from underground mining declined (Figure 6.3). Today, underground production is marked by low efficiency and high labour intensity, with output per man-shift from underground mines of CIL at 0.76 tonnes, against 12.3 tonnes from opencast mines. And the capital costs of underground mining are far higher.

Delays in clearances are another factor hampering coal production.[157] And CIL's manpower adequacy is a potential barrier to rapid and massive capacity expansion. The company has not recruited management personnel for several years and is becoming a company with executives whose average age is 50, with an attrition rate among them at 4% a year.

Figure 6.3.

Production of India's coal from opencast and underground mining (%)

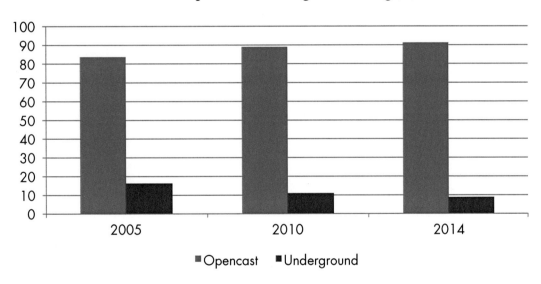

Source: Coal Controller's Organization 2014.

The shelf-life of a coal mine (underground or opencast) is about 30 years. If the government overcomes the foregoing hurdles, it will lead to a lock-in not only for capacity expansion but also in other ancillary services such as coal washing and transport. And if the government manages to install 175 GW of renewable capacity by 2022, too much investment in coal mining may waste resources. Instead, imports could make up the gap between demand and supply, which would also add to renewable capacity and help regulators bring parity to renewable and thermal-based power generation.

Sector governance

Policy paralysis after 2012, stemming chiefly from controversy surrounding awards of coal blocks, had a profound impact on the coal sector. The Comptroller and Auditor General's report of 2012 highlighted the problems. What were the issues exactly? What actions did the report prompt? And what was the cumulative effect on the sector?

The Coal Mines Nationalization Act of 1973 abolished private mining and brought the coal sector under the purview of the state and central governments. Subsequent amendments in 1993 and 1996 permitted captive coal mining in power, cement and iron and steel. These sectors were allotted coal blocks based on applications vetted by an inter-ministerial committee of the Ministry of Coal (MoC), CIL, Singareni Collieries Company Ltd. and members of ministries. But there was no transparency since the committee kept no records of the comparative evaluation and allocation process.

The 2012 report stated that the government had delayed introducing a competitive bidding auction despite approval from the Department of Legal Affairs in July 2006. The report also said a flawed process had led to a possible loss of Rs. 1.85 trillion to the exchequer, as the committee made the awards without estimating the fair price of coal available in the blocks.

The Central Vigilance Commission asked the Central Bureau of Investigation to look into the matter, and "public interest litigations" were filed in the Supreme Court, which allowed it to monitor the case. Amid accusations of corruption, the government set up an interministerial group to examine the coal block auctions case by case and to recommend ones to be de-allocated, causing the industry to "cry foul" and disrupting captive coal production. A 2014 Supreme Court judgment upheld the views of the Comptroller and Auditor General on the allocation process, stating that the blocks had been allocated illegally and arbitrarily, thus cancelling 204 of the 218 blocks. The new government accepted this judgment as well as the recommendation to re-auction the blocks, using a transparent auction-based process.

This led to the promulgation of the Coal Mines (Special Provisions) Ordinance 2014 (later, a bill), which described the mechanisms of a new auction process. Although such a method had been needed originally, the cancellation of the blocks violated the sanctity of contract of even the blocks that had been allotted fairly. It meant, too, that many companies, which already had sunk costs, would find it hard to take part because of pending bank guarantees. Finally, it meant that companies that had already built their end-use plant near the coal mines awarded to them would find it difficult to keep these plants running if they were not subsequently allotted the same mines.

Under the new process, the Nominated Authority (appointed by the government) would look into the eligibility of the each of the bids (technical, financial and compliance with other laws), which would be followed by two rounds of bidding. The regulated sector (power) would follow a reverse-bidding process in which the bidder that quotes the most discount to the notified price for a similar grade of coal wins the block. The nonregulated sectors (steel, cement and captive power plants) would follow the forward auction process in which the winner is the one who quotes the highest amount relative to the notified price for a similar grade (Ministry of Coal 2014). Moreover, for blocks reserved for the power sector, a fixed reserve price of Rs. 100 per tonne must be paid to the state government according to actual production above the royalty amount.

These auctions brought transparency to the process and increased accountability of the allottees, but raised certain questions over their effect on the sector as a whole. For example, the auctions for most blocks in the power sector began as reverse-bidding but ended up as forward-bidding since the bid price quoted by the bidders was zero. This essentially meant that the fuel cost (cost of mining) in a power purchase agreement would be borne by the winning bidder. In electricity tariffs, which mix fixed and variable costs, the fuel cost forms a major part of the variable cost, so that while consumers would benefit from lower electricity costs, producers could incur losses on fuel costs. For the winning bidders on a levelized basis over 25 years, the under-recovery in fuel cost is estimated to range from Rs. 0.39/kWh to Rs. 1.02/kWh, with aggregate under-recoveries estimated at Rs. 8 billion in 2015–2016 and Rs. 18 billion by 2017–2018. So, if power producers do not manage to increase fixed costs in power purchase agreements, this model could eventually lead to retrospective hikes in tariffs, once again violating the sanctity of contracts.

Another point of concern is that the government has made public only the final bidding price and not the entire data set. The argument is that the bids are part of a company's strategy and thus should remain confidential. But making the bids public would help in understanding the pattern and checking for discrepancies. Concerns over anticompetitive practices such as cartelization have arisen, with questions why the competitors could not keep lifting prices while still holding to the notified price limit. Following these allegations, the MoC rejected the bids from Jindal Power (Tara and Gare Palma blocks) and Balco (Gare Palma IV/I). Such moves once again create uncertainty: Are the auctions valid if the government can cancel the bids?

Governance issues have thus led to the perception of uncertainty in the coal sector, undermining coal supply. They jeopardized investments of Rs. 4 trillion in end-use plants, employment of almost 10 lakh people, and loans of Rs. 2.5 trillion by banks and financial institutions.[158] But the new auctions, which began by being more transparent than earlier procedures, have raised concerns over much-needed stability in the sector.

The need for an integrated policy environment

Given the complexity of the energy system and the federal structure of the government, it is vital to coordinate policy across all sectors. The coordination must extend not only to the energy supply and demand sectors but also across institutional structures, since an integrated energy policy can work only if it has these structures in an enabling environment that ensures efficiency and fairness.

A key aspect for policy is to make available the most appropriate technologies in the energy demand and supply sectors. The policy vision needs therefore to focus on:

- Setting a clear, transparent, long-term technology vision for the country, grounded in and supported by shorter term policies and plans.
- Enabling research and development (R&D) so that the country is in a state of readiness with an appropriate range of technologies to be harvested and deployed continuously over the long term.
- Putting forward programmes and measures that incentivize the use of efficient technologies by internalizing true costs and benefits (as much as possible). With affordability in mind, certain sections may still need targeted subsidies or "second-best" solutions until alternatives can be made available.
- Ensuring that policies can influence relative prices of alternatives appropriately for consumers, to provide the right signals and enable them to make rational energy choices.

Linking supply to demand

Industry

Industry's demand for energy is expected to grow sixfold between 2011 and 2050,[159] largely met by fossil fuels since industry relies so heavily on them. Much of the increase is expected to come from coal used for captive power generation by industries that have unreliable grid power and that can get cheaper power from captive plants. Given the negative externalities associated with coal-based power, and the imperative to move towards renewables, a dynamic policy should incorporate industry's fast-expanding power needs.

In the long term the government should provide a consistent policy that would enable the industry sector to move from captive power generation. Developing a smart grid that enables two-way interaction and increases reliability would be an important enabler for industry. Similar action is required in planning for oil and gas demand. (The fertilizer sector has shown a strong move from naphtha and towards natural gas; demand for fertilizers is expected to increase threefold by 2050. Given the paucity of natural gas it is important to consider how this demand would be serviced, especially in a politically sensitive and relatively inflexible sector like agriculture.)

The government has shown a strong commitment to mitigating the impact of climate change by monitoring energy efficiency, introducing the Perform-Achieve-Trade (PAT) under the National Mission on Enhanced Energy Efficiency. The PAT is an example of a well-planned and consistent policy, as it mandates the reduction of energy consumption by 8 mtoe in the first cycle, ending in 2015. It has been a success and should maintain momentum and include more sectors within it.

The Bureau of Energy Efficiency, in the Ministry of Power, was set up in March 2002 under the provisions of the Energy Conservation Act 2001, with the objective of reducing the energy intensity of the Indian economy by moving towards a self-regulated and efficient market-based system. The Bureau has been involved in creating strategies to facilitate measurement of energy conservation. In its regulatory capacity, it has developed energy efficiency codes for buildings and industries, certified professionals to conduct energy audits and developed norms for energy consumption. While much needs to be done to ensure complete coverage of energy efficiency standards in appliances, buildings, logistics and transport, among other areas, the activities of the Bureau are already positive steps in this direction.

Transport

In the reference scenario (BAU) of the TERI analysis, demand for petroleum is expected to increase 10-fold by 2050, doubling the country's import dependence. Transport in India has witnessed rapid expansion in use of personal vehicles fuelled by hydrocarbons. Apart from a focus on technology and infrastructure, the sector needs policy support to enable a modal shift to public transport and a fuel-switch for a smooth and gradual transition towards greater use of biofuels and electric vehicles.

The National Electricity Mobility Mission is promoting the uptake of electric vehicles through fiscal support, but without such associated infrastructure as charging stations, it will achieve little. For that reason the national Smart Grid Policy aims to incorporate a phased inclusion of charging infrastructure reflecting the stages in the Mission. The implementation of both these policies can be achieved only through concerted effort.

A forward-looking biofuel policy is important. Detailed policy guidelines have been provided for the first generation of biofuels, such as biodiesel from nonedible oil crops and bioethanol from sugarcane molasses. But the scope of expansion through these fuels is limited because of feedstock availability and large land requirements, which requires a shift towards more advanced biofuels. Lignocellulosic biofuels have been identified as potential options, and there has been mention of encouraging R&D and financial support. But the current policy mentions advanced biofuels as aspirational—a clear road map is needed, one to identify promising biofuel feed stocks relevant to India and the gaps in conversion technologies.

An intensive national biofuel R&D programme that creates awareness, encourages extensive R&D and has clearly defined goals needs to be put in place for a longer term solution. The alternative scenarios in RES explore a very ambitious biofuel uptake of as much as 40% of total transport demand in 2050; UCG explores the impact of meeting 14% of total transport demand with biofuels in 2050.[160] The key difference in the penetration of biofuels would be through greater policy direction and clarity.

Agriculture

Agriculture grew at a sluggish 4% a year from 2000 to 2015. With a large population dependent on the sector for their livelihood and given food security concerns, policies are needed to increase the sector's productivity and growth. Initiatives such as the National Mission on Agriculture Mechanization, which focuses on increasing the penetration of tractors, tillers and so on, are likely to increase energy requirements. It is hoped that investment in solar energy, irrigation and more profitable long-term plantations will be forthcoming with market reforms and the promise of a more stable and secure environment for small farmers.

The linkage of agriculture with water and other resources has prompted the government to craft many policies for electricity pricing, micro-irrigation, crop diversification and energy efficiency. But policies promising cheap power to farmers have merely created perverse incentives to grow water-intensive crops in areas already water stressed, aggravating water difficulties.

Just putting in place policies is not enough: ensuring that farmers can actually get the benefits of microfinance schemes and agricultural insurance schemes, by understanding their challenges, is vital to ensuring these initiatives' success. The stream of documents required for farmers to get these benefits can act as a primary obstacle, however.

Reform in agricultural pricing is another important element in farmers' financial inclusion. Access to free and fair markets, as well as policies that provide a sense of security, could automatically open channels for increased borrowing and technological progress, spurring growth. Today, however, the sheer number of policies means that governments must strive to avoid sending mixed signals.

Residential and commercial

India's urbanization is expected to increase swiftly. Many residential and commercial buildings are expected to be built—in fact, it is estimated that 70% of the buildings required in 2030 have yet to be built. A forward-looking policy can prevent lock-in into an energy-inefficient setup. A gradual move towards mandatory adoption of the Energy Conservation and Building Code should be considered. And because much of the urban infrastructure has yet to be put in place, proper urban planning and development of smart cities, if taken up now, would be striking the iron while it is hot.

The current framework for smart cities focuses on public transport, walking and cycling. With this in mind, it is important to plan for built-up structures to grow vertically in smart cities, primarily because horizontally developed cities require longer travel distances. Policies are needed for land development and water use; air quality standards; prices of alternative modes of transport in city centres; energy and appliance prices for consumers to make rational and efficient choices; incentives to green development; and integrated planning of water supply, waste management, the built-up environment and mobility planning in urban centres.

India's commercial sector is also growing fast owing to lifestyle changes and rising incomes. The number of shopping malls, hospitals and large office buildings is increasing rapidly, as is the share of air-conditioned commercial spaces. Given the high energy consumption tied to such development, and the stresses of new commercial and business districts on water availability and environmental quality, holistic urban planning is a prerequisite.

Although urbanization may reach close to 40% by 2050, a large part of the population will still be in villages. That makes it important to raise living standards in rural areas and prepare for smart villages along the lines of smart cities—to contain in-migration to city centres and to address energy access solutions, possibly through decentralized and renewable options for power.

Holistic planning and assessment of energy requirements are required from the demand and supply sides. And with the country's GDP expected to grow at around 7.5% a year, there is some urgency to tie up needs and developments, and plan accordingly.

Ensuring energy access

Ensuring energy access to all households is an essential aspect of an equitable energy system. So, making basic energy (fuels and services) available to even the remotest regions and rural settlements becomes imperative. And that requires listing and mapping household energy consumption patterns. Several studies have tried to do this, including a recent primary survey-based study, summarized here.

The high sensitivity of rural households to changes in fuel costs and the greater availability of alternative cleaner and efficient energy options place a burden on policymakers to enable household transitions to cleaner fuels through inclusive policies. Since the electricity transmission and distribution network has yet to cover every household in the country, and since households have yet to shift completely to LPG for cooking, the study sought to understand household perceptions of, and aspirations for, improved energy services. It indicated that the affordability of a fuel is a key determinant in the type of fuel chosen, highlighting the "ability to pay" (household income) and the "willingness to pay" (based on tastes and preferences, availability of the fuel and the opportunity cost of the fuel). Drawing on the results from more than 6,000 surveyed households across six states, Table 6.1 summarizes household preferences for alternative lighting and cooking fuels, and the costs that households were ready to bear.[161]

Table 6.1.

Willingness to pay for improved energy services

Maharashtra	Himachal Pradesh	Karnataka	Goa	Rajasthan	Odisha
Lighting device preferences (%)					
Compact fluorescent light bulb (%)					
71	51	43	52	40	54
Solar lantern					
20	36	50	9	40	16
Willingness to pay for lantern (Rs per month)					
Rent: 45 **Buy: 250-300**	Rent: 45 Buy: 450–800	Rent: 60 Buy: 250–600	Buy: 800–850	Rent: 90 Buy: 200–500	Rent: 120 Buy: 200–500
Cooking preferences and willingness to pay					
Improved cook-stoves (% willing to switch; maximum price in Rs)					
20; 800	14; 1,000	57; 600	55; 2,000	16; 800	30; 800
LPG connection (% willing to pay up to Rs. 1,500)					
66	63	73	54	51	80

Source: TERI 2014.

The larger number of households preferring CFL bulbs to solar lanterns as an energy-saving alternative can be attributed to the high initial cost of a lantern (Rs. 1,500 plus maintenance). The low willingness to pay for the lantern indicates the need for a financing mechanism that helps spread its cost over time (instalments, credit and so on). And since a solar lantern uses essentially no electricity and is easily portable, the benefits of the technology need to be marketed well, at times targeting certain household categories and in seasons with more days of sunshine. In short, mechanisms that make solar-based lighting devices easier to buy and use must be developed and implemented.

On the demand side, the costs associated with fuel use heavily influence individual energy choices. For cooking, LPG certainly stands as a cleaner and more efficient fuel, but low LPG coverage in rural areas points to the high costs of getting a connection. The survey reveals a willingness to pay of up to Rs. 1,500 for an LPG connection—against a current LPG connection charge of up to Rs. 4,000. So, to raise the uptake of LPG in rural households, subsidies could be designed to cover part of the connection cost above that willingness limit. A reallocation of unused resources from cylinder-based subsidies, and savings from PAHAL (the Direct Cash Transfer Scheme to subsidize new LPG connections), could be considered.

LPG use in rural households is also held back by supply-side constraints of inefficient delivery and supply. The added transport cost—cylinder delivery outlets are often more than 8–10 km away—dissuades households from completely shifting to LPG for cooking.[162] Widening the distribution network and designing benefit-sharing delivery mechanisms could further improve access to LPG cylinders.

Households with intermittent or no electricity access rely on kerosene lamps for lighting. Many households also use modest quantities of kerosene for cooking, augmenting it with biomass fuels. But kerosene, a subsidized fuel, is susceptible to excessive leakage into the black market and to adulteration with transport fuels, limiting the availability to households. Efforts to plug such leakage and to reform the subsidy mechanism were initiated through the Direct Benefits Transfer Scheme, launched on a pilot scale for kerosene in Alwar, Rajasthan. A direct transfer of subsidies into the beneficiaries' bank accounts meant that the fuel was available at market rates. With the benefit of the difference between the market and subsidized price no longer present for commercial entities, kerosene uptake declined sharply. Since kerosene is sold at village ration shops, expanding this scheme to states with high kerosene use would not only ensure the fuel's optimal use in households but also ensure subsidy savings, which could then be directed to developing possible decentralized energy options.

Generating thermal power

The power sector is the key energy sector—as a producer and consumer of energy, as a strong link to economic growth and as a provider of energy to all sections of society. Its environmental and resource-use implications are also notable and so deserve full attention from the policy perspective. For India, power generation is going to be led mainly by the intent to provide enough power—affordably. The economic viability of power supply options over the entire energy chain (including extraction, conversion and distribution) will remain a basic decision-block in adopting technologies.

Given the coal-dominant base that the sector is likely to retain over the next two decades, several issues need to be faced. Where should the coal be sourced, and how? How will power generation technology progress over the years? How much capacity should be developed for coal transport and washing? What levels of port capacity will be required for coal imports? The list goes on. Simultaneously, as renewable technologies mature further and relative costs of alternative technologies become more attractive, power should be ready to take up these options. One of the key roles of an integrated energy policy is to enable the adoption of technologies compatible with India's energy portfolio, based on the relative costs to end-users.

Power tariffs have fundamental political underpinnings, influencing key decisions in generation, transmission and distribution. Characterized by financially distressed utilities, low electrification rates and unreliable supplies, the sector has to re-evaluate pricing mechanisms and carefully consider the fuel choices.

Policy directions on investing in power-generation technology need to be informed by relative costs: the landed cost of underground versus imported coal, and the cost of power from domestic or imported coal. The following quantitative analysis drills down into those costs.

Underground or imported coal

A push towards underground coal mining in India should be evaluated against the possible use of imported coal. So far, the higher costs of imported coal have been a major deterrent to promoting its use. Based on a TERI analysis of relative costs of imported coal and underground domestic coal for the 1,200 MW Dahej Power Plant, imported coal could be much cheaper than underground coal (Rs. 5,731 a ton vs 6,052).[163]

Similarly, externalities linked to coal extraction and use, if internalized into policymaking correctly, could heavily influence the relative costs of alternative sources of generation. Policymakers must also consider the relative benefits of investing in new and additional washing capacity for coal, in light of the imminent increase in other clean or renewable-based capacities.

While geopolitical stresses in international coal markets could lead to price volatility, importing coal could be a transition strategy in moving towards a sustainable and renewable energy future.

T&D infrastructure and grid stability

Rising capacity for power generation has not been matched by a proportionate increase in that for power transmission, and greater electricity demand from consuming sectors has put more pressure on regional grids. But expanding transmission capacity is not a short-term matter: it is slowed by long gestation periods, by inordinate delays in acquiring land and official clearances and by right-of-way issues. These obstacles have also curtailed the evacuation of excess power to power-deficit regions.

In response to high renewable energy targets and the need for shifting towards electric vehicles, a comprehensive energy policy must consider technological shifts (such as moving from low-voltage lines to high-voltage distribution systems—Chapter 3), institutional barriers (harmonizing policies among ministries) and governance mechanisms (minimizing power theft). To manage the fast-growing power system, investment is needed to build a smarter system, incorporating information-technology, communication and automation systems.

The Smart Grid Roadmap for India has a comprehensive plan to enable such a national transition. But it does not provide state or utility plans, where most of the work is required. Another important part of the Smart Grid Policy that does find mention in the Roadmap but requires a deeper analysis is the use of low-cost domestically produced smart meters and a detailed plan for implementing them at state and utility level. Investments, skilled human resources and good customer outreach and communication will all be vital for rolling out the policy.

LNG terminals and natural gas pipelines

Natural gas is likely to be a key element in developing the energy sector. Owing to domestic shortages and uncertainties of cross-border pipelines, LNG imports are important in the short term.

To ensure that the natural gas market in India is robust, one of the primary requirements would be an extensive pipeline network (Figure 6.4). The newest terminal at Kochi was commissioned in January 2014. The use of terminals varies. While the Dahej Terminal operated at 110% of its nameplate capacity, the Kochi terminal was greatly underused, primarily because of delays in building pipeline infrastructure. Underuse of some existing pipelines and delays in signing supply agreements and end-user gas purchase agreements have been some of the checks on rapid development.

With the recent drop in spot LNG prices in Asia, the emergence of new suppliers (Australia, the United States and possibly Russia among others) could lead to a more dynamic and price-competitive LNG market for India. Beyond reforms to policies on supply and pricing for natural gas demand (primarily fertilizer and power), the infrastructure requirements for gas supply also need to be tackled—and urgently. At the August 2014 budget session the government stated its intention to develop 15,000 kilometres of additional pipelines.

Strengthening institutional links

Energy's institutional structure is quite complex, with multiple institutions at state and central levels and with overlapping areas of jurisdiction (Figure 6.5). While natural to a federal structure, this complexity impedes efficiency. The public sector generation companies are under the central government, while the distribution companies (DISCOMs) are under the states. Therefore, the generation companies would pass on an increase in fuel prices to the DISCOMs, but because the DISCOMs are much closer to the state, they cannot pass the price increase on to the consumers. This leads to an inefficient structure where DISCOMs bear the losses, which are in turn passed on to the exchequer. In addition, the constitutional arrangement for resources leaves disproportionate burdens on the state and central governments. For instance, in setting up special export zones for manufacturing, the state often has to forgo duties and taxes and to provide R&D, while the central government usually provides the mandate for acquisitions of land.

Figure 6.4.

Existing and planned LNG terminals and natural gas pipelines

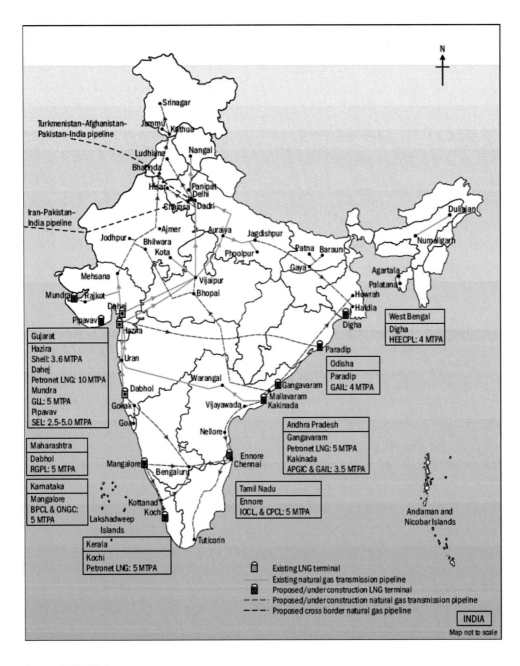

Source: TERI 2015.
This map is not to scale and does not depict authentic boundaries.

Figure 6.5.

Main public sector actors in the energy sector

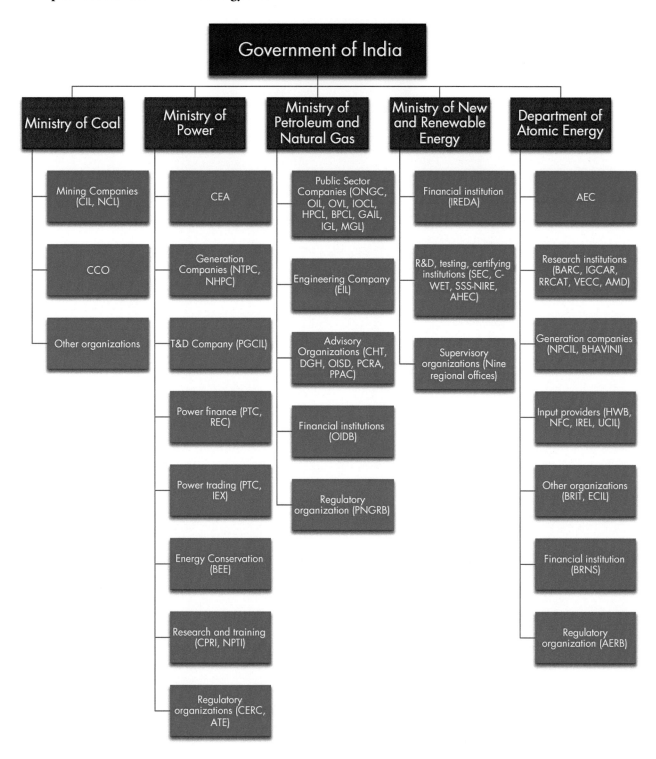

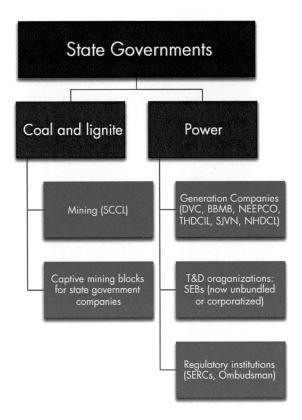

Source: TERI 2015.

With existing capacity of a mere 3 GW as of March 2015, the central government revised capacity-addition targets for solar energy from 20 GW to 100 GW by 2022. It also raised overall renewable energy generation targets to 175 GW for that year. The new targets are ambitious and will at a minimum call for policy harmonization, rapid project clearances and good access to finance. Today, however, the sector faces a raft of impediments, including multiple implementing agencies, hurdles in land acquisition, policy uncertainty over renewable purchase obligations (RPOs) and inadequate long-term finance. Harmonizing policies between central and state governments is essential for the success of this programme.

Although the central government has proposed overarching national targets for renewable energy, implementation will largely be at state level. The issue becomes particularly acute with RPOs, introduced in 2010 under the National Tariff Policy. By 2020, 15% of the total power requirement is to come from renewable sources. However, these RPO targets—set by the central government—are not being adequately enforced at state level, possibly compromising medium-term national targets.

Differentiating the relative attractiveness of fuel and technological options at different locations needs to be included in policy prescriptions and decision-making criteria. In addition to the 175 GW target, the government has targeted 1.5 billion tonnes of coal production by 2022. This seems important for increasing energy access in the country. But given its relative competitiveness coal could divert investment from renewables, unless externalities are internalized into the cost of generation. Moreover, this may also mean over investment in energy supply in the short term, but imply a long-term lock-in as well. Accordingly, to be in sync with other goals—like improving air quality and reducing dependence on fossil fuel imports—policies should move towards imposing appropriate costs on environmental degradation and the like. Finally, plans for the large-scale integration of renewables in the grid need to address intermittency associated with the variable supply of renewables.

All these intricacies underline the need for an efficient energy-transformation system and for well-demarcated domains of jurisdiction among the institutions in the energy sector—and thus for a strong energy planning and enabling body.

Moving to independent regulation

India needs an independent regulator, one that can help set up and oversee liberal markets and protect laws from populist forces and political pressures. The regulators must not only be financially and statutorily independent—they must be seen as such. Since the central government is a policymaker with interests in commercial activities through public entities, an independent regulator could help ensure that neither public nor private companies unduly benefit.

In theory, some of the crucial requirements of such a regulator are a clear legal framework and legal powers—statutory authority; clarity of roles; transparency; predictability and flexibility; financial independence; and enforcement and dispute-settlement authority. Indeed, independent regulation is necessary for liberalizing the market.

Several instances show the need for a strong independent regulator. In oil and gas, the technical arm of the Ministry of Petroleum and Natural Gas (MoPNG)—the Directorate General of Hydrocarbons (DGH)—was formed as a technical regulatory body for the upstream segment. The resolution passed to constitute this states that the DGH should "promote sound management of the Indian petroleum and natural gas resources having a balanced regard for the environment, safety, technological and economic aspects of the petroleum activity."

While the DGH is in essence responsible for developing the sector, the resolution emphasizes its activity as an advisory body to the government, administratively coming under the MoPNG, while its lack of statutory and financial independence reduces its effectiveness. Consultations with stakeholders echo the recommendations of various expert committees (such as the Chawla Committee and the Kelkar Committee) on the need for independent upstream regulation. But the MoPNG has contested this on the grounds that, since the government is the owner of natural resources, it has an important role in managing and developing them.[164] Also according to the MoPNG, the NELP and the Coal Bed Methane policy already provide a level playing field. And since a PSC is between the government and the contractor, independent regulation "may not be tenable." In short, there has been no resolution of the DGH's role in developing the sector. Moreover, DHG officials are appointed on the basis of deputation from the oil companies, whose activities the DGH is supposed to govern—a revolving door that could produce conflicts of interests.[165]

The midstream and downstream segment's regulator, the Petroleum and Natural Gas Regulatory Board (PNGRB), is mandated "to regulate the refining, processing, storage, transport, distribution, marketing and sale of petroleum, petroleum products and natural gas excluding production of crude oil and natural gas." It purports "to protect the interests of consumers and entities engaged in specified activities relating to petroleum, petroleum products and natural gas." The objective is to "ensure uninterrupted and adequate supply of petroleum, petroleum products and natural gas in all parts of the country and promote competitive markets and for matters connected therewith or incidental thereto."[166] Companies need to register with the Board for operating LNG terminals as well as establish storage facilities beyond capacity. The PNGRB, also a dispute-settlement institution, has the same powers as the civil courts, which make it a more effective regulator than the DGH.

Even then, some constraints prevent the PNGRB from regulating optimally. It has little control over deciding the location of refineries and does not have a mandate to grant operating licenses. Government involvement in appointing board members also raises questions of independence. Nor is there clarity on the purview of its authority. For example, Indraprastha Gas Ltd. challenged the PNGRB's authority to regulate tariffs and compression charges—and the jurisdictions of the Competition Commission of India and the PNGRB.

The oil sector also has the Oil Industry Safety Directorate, to improve safety standards, and the Oil Industry Development Board, to promote and facilitate sector development. The latter collects a cess on blocks awarded on a nomination basis and extends loans to companies. Adding the Directorate General of Mines Safety and the Petroleum Explosives Safety Organisation to the list of regulators with some authority in the sector complicates the picture even further.

For coal, regulations are managed by the Ministry of Coal (MoC), the Coal Controller's Office (CCO) and CIL. Of the three the CCO has only supervisory powers for collecting data, so the MoC and CIL decide and influence changes in the market. Once again, with the government as both policymaker and producer (through its roughly 90% share in CIL), the need for an independent regulator did not arise. But now the government is moving to liberalize the sector. In 2015 the legislature passed the Coal Mines (Special Provision) Bill, which allows commercial mining.[167] That introduces competition to a sector that until now was closed and controlled and thus creates a need for an independent regulator to ensure a level playing field for all coal producers.

This need is especially evident in coal pricing, coal exploration and dispute resolution. Coal pricing, deregulated in 2000, is decided by CIL based on gross calorific value grades, but the lack of a dynamic coal market (where CIL was the primary supplier) makes pricing highly uncertain and opaque. With private players now entering the market, the basis for pricing has to be set by an unbiased body.

Coal exploration has so far been under the purview of government agencies such as the Central Mine Planning and Design Institute (a subsidiary of CIL), the Geological Survey of India and the Mineral Exploration Corporation Ltd. Thus CIL, through the Institute, has an inordinate amount of influence even in the exploration of blocks and the allocation of mining plans, which should be curtailed by an independent body.

Finally, disputes between power generation companies and CIL over the quality of coal have arisen many times in the past, with the producers alleging that CIL is abusing its monopoly position to deliver a lower grade of coal than agreed to in the fuel supply agreement. Without a redressal mechanism, such issues are brought before the Competition Commission of India, which lacks the teeth for this work.

The government had in fact by 2013 initiated plans to set up a regulator on the lines of the DGH. Under the Coal Regulatory Authority Bill 2013 the authority would have specified methods for declaring grades; monitored or enforced mine closure; ensured that the mine developer adhered to the approved mining plan; adjudicated disputes between parties; called for information or published data on the coal industry; and advised the government on policy formulation (including allotting or earmarking coal blocks for any purpose or mode). The final bill, however, removed the power to decide prices, and the regulator could only suggest methods to the coal companies. This bill lapsed with the dissolution of the 15th Lok Sabha. But there have been reports that the government is considering re-introducing a coal sector regulator bill, which will help to determine a mechanism to price coal and specify methods to ensure supply of high-quality coal.[168]

The nuclear sector stands somewhat apart, but has similar issues. The Atomic Energy Regulatory Board falls under the Department of Atomic Energy but also receives financial support from it. The Comptroller and Auditor General (2012) has even questioned the Board's legal status. Another issue is the system of revolving doors. As military and civil use of nuclear power are separated,[169] more players are likely to enter the sector (Rosatom and ongoing discussions with AREVA and Westinghouse are cases in point). The possibility of joint ventures between the Nuclear Power Corporation of India and domestic private and public companies is also being explored.

To avoid the possibility of developing bias, to address safety concerns and to bolster investor and public confidence requires a regulator with strong statutory powers. The Nuclear Safety Regulatory Authority Bill, presented to Parliament, aims to achieve this.

Conclusion

India needs an integrated, stable and consistent energy policy to pursue its long-term vision. The policy has to address links between demand and supply among the energy sectors and among the economy's non-energy sectors. It should also fully reflect the relative costs of alternative energy choices for consumers. And it should ensure reliable and consistent supplies of energy over the long term, encompassing elements of resource sustainability into the planning and policymaking architecture.

Economywide, the policy should incorporate a robust framework of demand-side assessment and management for forecasting energy needs and allow for transparent and coordinated planning. It also has to manage the institutional structure for energy's smooth transition to more liberal markets, ensuring a good investment climate and offering a clear long-term direction for stakeholders. Hence the call for an integrated policy that assesses needs across sectors and provides a coherent institutional structure and the requisite enabling frameworks. In formulating such a policy, transparency and stakeholder engagement are essential. The overall policy and regulatory system has to build trust so that it can take tough decisions to reconcile often-conflicting challenges.

APPENDICES

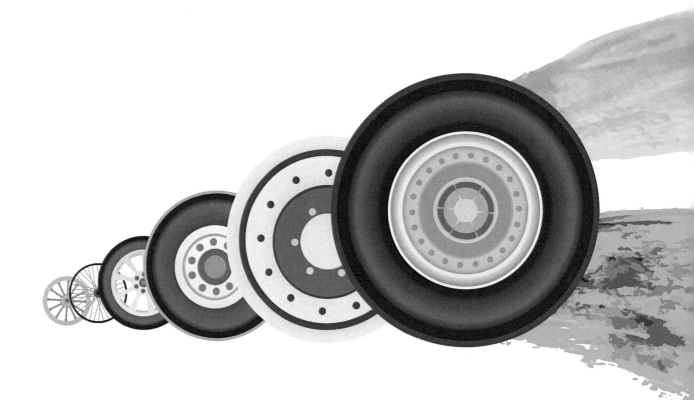

Appendix 1

India Energy Model (By CEEW)

Brief Description of India Energy Model

The India Energy Model is a modified/ adapted version of GMOS/NetSim, a supply chain planning tool developed by Shell Global Solutions International. GMOS/NetSim is used for various supply chain challenges ranging from studies concerning strategic investments decisions to fact-based decision support over short and long planning cycles. The key strength of GMOS/NetSim is that, due to its generic building block set-up, it can be used to model a wide range of supply chain challenges. The underlying mathematical model—using AIMMS as the modelling language is solved as a linear programming problem. In the case of INEM, the tool is used to arrive at a least cost energy system for the future based on a perfect foresight on costs of technologies and cost of fuel resources and demand structure from the various sectors of the economy. The model is run over a 40 year time horizon (2011–2050) in investment time steps of five years to capture the evolution of energy system.

The Scenarios

The two long-term energy scenarios evaluated using INEM are the Wait and Watch (W&W) scenario and the Low Carbon Inclusive Growth (LCIG) scenario. As the name alludes, W&W refers to a scenario where developments in the energy system progress as per current policies, albeit with an awareness of the impact of unbridled energy consumption. There is a sustained interest in furthering the role of coal and also the roll-out of RE to the extent that it is needed to bridge the gap that fossil fuel-based sources leave. Ending energy poverty and improving access to commercial fuels in rural areas is already a priority in this scenario. Autonomous energy efficiency gains are seen across the industrial sector, as envisaged in current schemes such as PAT and the extensions thereof. Building energy efficiency also improves in this scenario as there are regulations in place and wide-spread acknowledgement of the benefits are expected in due course. Sectors such as transport and agriculture do not see a concerted shift away from the dependence on fossil fuels. There are no constraints imposed on economy wide CO_2 emissions in such a scenario.

The LCIG scenario has some similarity to the W&W scenario in sectors where concerted action is already underway. Industries, residential and commercial sectors exhibit high levels of efficiency. In addition fuel switch to efficient fuels is incentivized in the industrial sector. There is a significant shift in the power sector mix where aggressive roll-out of RE sources is a policy target. No explicit targets are set but the ability to ramp up capacity over time is less constrained in this scenario (seen as an outcome of a specific policy to promote RE). Transport and agriculture sectors, again do not have specific policies for a move to cleaner technologies. The overall constraint on the CO_2 emissions from the energy system provide significant incentive for industrial fuel switches, power sector transition and for a move to cleaner technologies in transport and agricultural sectors.

The growth rates assumed in the exercise are not as high as those suggested in government communications. Instead, a tapering growth rate that starts at 8% in the short term and reduces progressively to 6% by the middle of the century forms an important assumption in the scenario analysis. The overall economic growth that is witnessed in both scenarios is of a similar magnitude. This implies that economic growth is exogenous to the nature of the energy system and drives the overall end-use service and product demand. As a result, the overall primary energy supplied to the economy is lower only by a mere 9% in the LCIG scenario. A bulk of the difference is on account of the decarbonization of the power sector. This suggests that much of the efficiency gains are captured even in the W&W scenario. In this sense the W&W is a departure from the conventional BAU scenarios that reflect an inefficient energy system under current policies and outlook.

Some results from the two scenarios are detailed below:

Overall Primary energy consumption in W&W (2035)—63.1 EJ, Fossil Share in TPES—91%
Overall Primary energy consumption in LCIG (2035)—61.4 EJ, Fossil Share in TPES—89%

Overall Primary energy consumption in W&W (2050)—98 EJ, Fossil Share in TPES—92%
Overall Primary energy consumption in LCIG (2050)—89.7 EJ, Fossil Share in TPES—83%

Power Sector (2035)

W&W: Fossil-based generation—89%, RE—4%, Hydro + Nuclear—7%
LCIG: Fossil-based generation—81%, RE—12%, Hydro + Nuclear—7%

Power Sector (2050)

W&W: Fossil-based generation—92%, RE—4%, Hydro + Nuclear—4%
LCIG: Fossil-based generation—68%, RE—21%, Hydro + Nuclear—11%

Industry (2035)

W&W: Electricity contribution to Industry end-use needs—19.6%
LCIG: Electricity contribution to Industry end-use needs—21.4%

Industry (2050)

W&W: Electricity contribution to Industry end-use needs—21%
LCIG: Electricity contribution to Industry end-use needs—31%

Shell's New Lens Scenario Insight

Shell's New Lens Scenarios paint two different pictures for India, however many of the issues that are raised by the two scenario worlds are common to both. Import dependency, reliance on fossil fuels, rising prices and access to energy are all issues that are witnessed in both Oceans and Mountains.

CAGR GDP/Capita	2000–2005	2005–2010	2010–2015	2015–2020	2020–2025	2025–2030	2030–2035
Oceans	3.4%	5.1%	4.4%	7.3%	5.3%	2.8%	2.7%
Mountains	3.4%	5.1%	4.5%	6.3%	5.0%	2.0%	0.2%

Assuming Indian economic growth follows the path set out in the table above, current NLS projections estimate that Indian total primary energy demand could see an increase of around 140% to 200% by 2035, which is an average of between 1.7EJ and 2.4EJ increase per year in the outlook period. Both scenarios see coal remaining the backbone of the energy system, though to varying degrees. By 2035, it could potentially still provide between 30% and 55% of primary energy demand, meanwhile gas could struggle in the Oceans scenario, supplying just 7% of demand compared to up to 20% in Mountains. The continued draw on coal in Oceans may be driven by the need of an energy-hungry economy to draw on all sources of fuel with the nearest (and cheapest) to hand providing the easiest option. The increased role for gas in Mountains is based on ruling powers being able to install policies that preserve their power, mostly using levers of the state to shape and steer social and economic policy. One particular beneficiary is unconventional gas, which despite still being unsuccessful in India, flourishes worldwide, leading to larger volumes (at moderate prices) of LNG being available. This could lead to concerted efforts to put in place infrastructure to enable gas to break into additional markets, and the more moderate pace of economic growth could buy enough time to construct gas-fired power stations, stealing market share of power generation from coal.

In a global context, Oceans could see India becoming the joint second largest energy consumer in the world (tied with the USA), making up around 11% of global demand. This is still some way behind China, responsible for a 24% share of demand. However, this 11% is significantly greater than the 5.4% share it held in 2010. This larger share, as already mentioned, is driven in Oceans by coal, which in turn would make India the world's second largest consumer of coal by the end of the outlook period. It could account for up to 25% of global coal demand, second only to the 40% of China. In this respect, India may well end up being a price setter for international coal markets, especially if its domestic production fails to reach potential.

Total Primary Energy Demand—Mountains

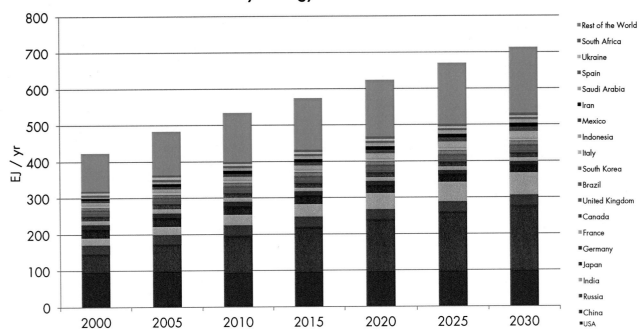

Total Primary Energy Demand—Oceans

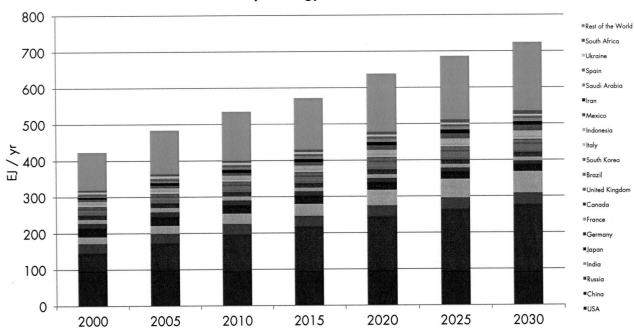

In Mountains, India would likely emerge as the world's 3rd largest energy market by 2035, behind that of China and the USA, not quite doubling its share, but nevertheless almost hitting 9%. Despite relying heavily on gas in this scenario, India never reaches more than 6% of global gas demand, behind China, USA and Russia. In this sense, India would be more of a price taker in gas markets. In Oceans, greater demand for oil could place India alongside the USA in second place of world oil consumers, though of course the US are blessed with greater indigenous resources, acting as a hedge against fluctuations in oil price. It is a similar story in Mountains, where despite oil consumption being lower, India could still become the world's third largest market for the fuel by 2035. It is currently fourth with a share of around 4%. This could reach up to 11% by 2035.

In an Oceans world, where global and Indian economic growth continues to be strong (based on an increasingly vocal population challenging vested interests and driving new waves of growth), diffusion of power downwards and the presence of more opposition to technologies such as unconventional gas and nuclear frustrate attempts at their development. This subsequent tightness between supply and demand would subsequently keep prices, in particular oil, high. Despite Mountains being a scenario with looser global supply–demand balances and more moderate price outlooks, oil price continues to rise in this outlook too.

This means that, coupled with a 3 to 3.5 times increase in oil demand, import dependency would rise in both scenarios, with implications for percentage of GDP spent on energy imports. Volumetrically, with oil import dependency possibly reaching almost 90%, would India consider putting import caps in place? Would sourcing of truly secure supply be the optimal solution, or are demand side measures the answer to relentless increases in oil demand?

The consumption of energy in the residential sector is another area that has been identified as being of interest for further research. Both scenarios have different stories for this sector. Slowing economic growth and its knock-on effect of reduced income per capita could see traditional biomass keeping a large share of the market, though being used increasingly efficiently due to the effects of incrementally better technology such as cleaner-burning cookstoves being installed. It may also transpire that stronger economic growth and higher energy prices lead to greater proliferation of distributed solar photovoltaics, with their off-grid capabilities also being attractive in a world of fewer large infrastructure projects and faltering grid improvements.

Both scenarios see passenger road transport consumption increase threefold. As the number of cars on the roads increases, so too will demand for fuels to power them. The scenarios illustrate different energy pathways that mobility could take. Could it capitalize on moderate gas prices and increase the number of gas-fired vehicles (as seen in the adjacent Mountains chart)? Could overall demand for cars be reduced by building smart cities, an environment far more conducive to the electric car? Or will it simply continue relying on liquid hydrocarbon fuels, countering rising prices by installing more efficient internal combustion engines and increasing the use of biofuels?

Should the rapid economic growth of recent times continue, this would place huge strains on energy sources, and despite aggressive renewables and efficiency policies, large volumes of fossil fuels would still be required to cover the majority of demand. This is the case in both scenarios, but the mix of fossil fuels is one that bears scrutiny. Would abundant natural gas (and therefore moderate LNG prices) ease security of supply fears, or alternatively might higher global gas prices encourage more investment into the domestic coal sector?

As for CO_2 emissions, rapid, coal-fired growth in Oceans could see a continuous increase in emissions up until at least 2035, as seen in the adjacent chart. Meanwhile in Mountains this could show signs of being on a plateau by 2035, thanks to CCS and gas-for-coal substitution. Might renewables spread faster in Oceans, replacing greater volumes of coal in power? Or could the cleaner-burning gas of Mountains be used in Oceans to compliment greater renewables and dramatically cut emissions?

TERI's MARKAL Model for India

MARKAL is a bottom up dynamic linear programming modelling framework. MARKAL depicts both the energy supply and demand sides of the energy system. It provides policymakers and planners in the public and private sectors with extensive details on energy producing and consuming technologies, and an understanding of the interplay between the various fuel and technology choices for given sectoral end-use demands. As a result, this modelling framework has been widely used for the development of carbon mitigation strategies. The MARKAL family of models is unique, with applications in a wide variety of settings and global technical support from the international research community.

MARKAL interconnects the conversion and consumption of energy carriers. This user-defined network includes all energy carriers involved in primary supplies (e.g., mining, petroleum extraction, etc.), conversion and processing (e.g., power plants, refineries, etc.), and end-use demand for energy services (e.g., automobiles, residential space conditioning, etc.) that may be disaggregated by sector (i.e., residential, manufacturing, transportation and commercial) and by specific functions within a sector (e.g., residential air conditioning, lighting, water heating, etc.). The optimization routine used in the model's solution selects from each of the sources, energy carriers and transformation technologies to produce the least-cost solution, subject to a variety of constraints. The user defines technology costs, technical characteristics (e.g., conversion efficiencies) and energy service demands.

As a result of this integrated approach, supply-side technologies are matched to energy service demands. Some uses of MARKAL include:

1. Identifying least-cost energy systems and investment strategies;
2. Identifying cost-effective responses to restrictions on environmental emissions and wastes under the principles of sustained development;
3. Evaluating new technologies and priorities for research and development.
4. Performing prospective analysis of long-term energy balances under different scenarios.
5. Examining reference and alternative scenarios in terms of the variations in overall costs, fuel use and associated emissions.

The MARKAL framework is detailed out in Figure 1.

MARKAL framework overview

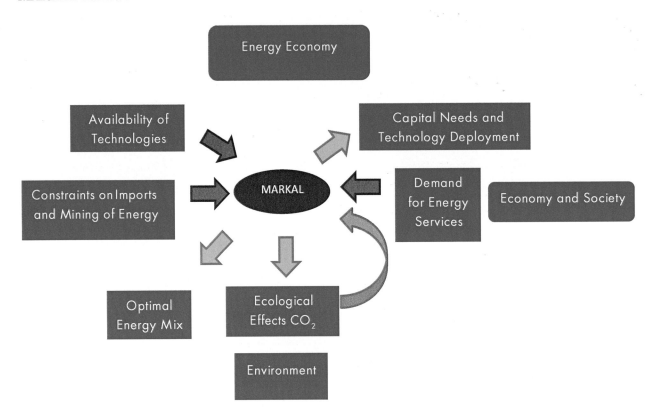

Source: IEA, ETSAP

The MARKAL database for this exercise has been set up over a 50 year period extending from 2001–2050 at five-yearly intervals coinciding with the Government of India's Five-Year plans. The year 2001–2002 is chosen as the base year as it coincides with the first year of Government of India's Tenth Five Year Plan (2001/2002–2006/2007). In the model, the Indian energy sector is disaggregated into five major energy consuming sectors, namely, agriculture, commercial, industry, residential and transport sectors. Each of these sectors is further disaggregated to reflect the sectoral end-use demands. The model is driven by the demands on the end-use side. End-use service level demands at the country level are estimated econometrically across various energy consuming sectors of the Indian economy.

On the supply side, the model considers the various energy resources that are available both domestically and from abroad for meeting various end-use demands. These include both the conventional energy sources such as coal, oil, natural gas and nuclear as well as the renewable energy sources such as hydro, wind, solar, biomass, etc. The availability of each of these fuels is represented by constraints on the supply side.

The relative energy prices of various forms and source of fuels dictate the choice of fuels which play an integral role in capturing inter-fuel and inter-factor substitution within the model. Furthermore, various conversion and process technologies characterized by their respective investment costs, operating and maintenance costs, technical efficiency, life, etc., that meet the sectoral end-use demands are also incorporated in the model. A discount rate of 10% has been assumed through this period. Prices of conventional fuels have been taken from fuel price projection published by the International Energy Agency's (IEA).[170]

India-specific capital costs and O&M costs for various technologies included in the database have been obtained from various sources. Wherever India-specific costs are not available, international figures are used. Cost reduction in future in the emerging technologies has also been assumed based on an understanding of the particular technology development.

The model structure is described in Figure 2.

Figure 02:
Model logic

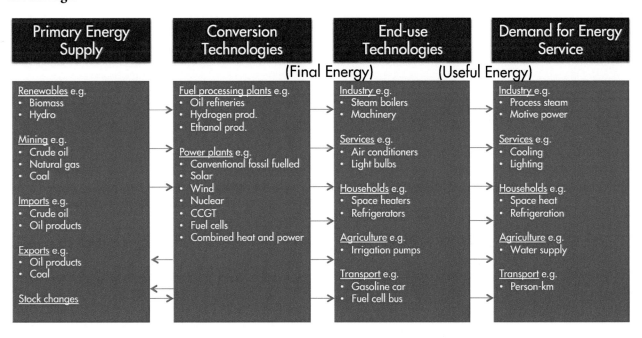

Source: IEA, ETSAP.

Alternative scenarios such as renewable energy scenario (RES) and the increased domestic production (UCG) were set up against the business as usual (BAU) scenario to examine different outcomes in terms of fuel mix, technology deployment, commercial energy requirement and investment.

Appendix 2

Cautionary Note

The New Lens Scenarios are part of an ongoing process used in shell for 40 years to challenge executives' perspectives on the future business environment. We base them on plausible assumptions and quantification, and they are designed to stretch management to consider even events that may be only remotely possible. Scenarios, therefore, are not intended to be predictions of likely future events or outcomes and investors should not rely on them when making an investment decision with regard to Royal Dutch Shell plc securities.

The companies in which Royal Dutch Shell plc directly and indirectly owns investments are separate entities. In this book "Shell", "Shell group" and "Royal Dutch Shell" are sometimes used for convenience where references are made to Royal Dutch Shell plc and its subsidiaries in general. Likewise, the words "we", "us" and "our" are also used to refer to subsidiaries in general or to those who work for them. These expressions are also used where no useful purpose is served by identifying the particular company or companies. "Subsidiaries", "Shell subsidiaries" and "Shell companies" as used in this book refer to companies over which Royal Dutch Shell plc either directly or indirectly has control. Companies over which Shell has joint control are generally referred to as "joint ventures" and companies over which Shell has significant influence but neither control nor joint control are referred to as "associates". In this book joint ventures and associates may also be referred to as "equity-accounted investments". The term "Shell interest" is used for convenience to indicate the direct and/or indirect ownership interest held by Shell in a venture, partnership or company, after exclusion of all third-party interest.

This book contains forward-looking statements concerning the financial condition, results of operations and businesses of Royal Dutch Shell. All statements other than statements of historical fact are, or may be deemed to be, forward-looking statements. Forward-looking statements are statements of future expectations that are based on management's current expectations and assumptions and involve known and unknown risks and uncertainties that could cause actual results, performance or events to differ materially from those expressed or implied in these statements. Forward-looking statements include, among other things, statements concerning the potential exposure of Royal Dutch Shell to market risks and statements expressing management's expectations, beliefs, estimates, forecasts, projections and assumptions. These forward-looking statements are identified by their use of terms and phrases such as "anticipate", "believe", "could", "estimate", "expect", "goals", "intend", "may", "objectives", "outlook", "plan", "probably", "project", "risks", "schedule", "seek", "should", "target", "will" and similar terms and phrases. There are a number of factors that could affect the future operations of Royal Dutch Shell and could cause those results to differ materially from those expressed in the forward-looking statements included in this book, including (without limitation): (a) price fluctuations in crude oil and natural gas; (b) changes in demand for Shell's products; (c) currency fluctuations; (d) drilling and production results; (e) reserves estimates; (f) loss of market share and industry competition; (g) environmental and physical risks; (h) risks associated with the identification of suitable potential acquisition properties and targets, and successful negotiation and completion of such transactions; (i) the risk of doing business in developing countries and countries subject to international sanctions; (j) legislative, fiscal and regulatory developments including regulatory measures addressing climate change; (k) economic and financial market conditions in various countries and regions; (l) political risks, including the risks of expropriation and renegotiation of the terms of contracts with governmental entities, delays or advancements in the approval of projects and delays in the reimbursement for shared costs; and (m) changes in trading conditions. All forward-looking statements contained in this book are expressly qualified in their entirety by the cautionary statements contained or referred to in this section. Readers should not place undue reliance on forward-looking statements. Additional risk factors that may affect future results are contained in Royal Dutch Shell's 20-F for the year ended December 31, 2015 (available at www.shell.com/investor and www.sec.gov). These risk factors also expressly qualify all forward looking statements contained in this book and should be considered by the reader. Each forward-looking statement speaks only as of the date of this book (18 March, 2016). Neither Royal Dutch Shell plc nor any of its subsidiaries undertake any obligation to publicly update or revise any forward-looking statement as a result of new information, future events or other information. In light of these risks, results could differ materially from those stated, implied or inferred from the forward-looking statements contained in this book.

We may have used certain terms, such as resources, in this book that United States Securities and Exchange Commission (SEC) strictly prohibits us from including in our filings with the SEC. US Investors are urged to consider closely the disclosure in our Form 20-F, File No 1-32575, available on the SEC website www.sec.gov.

Endnotes

1. Census 2011.

2. Population Foundation of India.

3. TERI data.

4. Government of India 2015 http://www4.unfccc.int/submissions/INDC/Published%20Documents/India/1/INDIA%20INDC%20TO%20UNFCCC.pdf (page 29).

5. http://blogs.shell.com/climatechange/2014/05/twosides/

6. Arunabha Ghosh (2015) "The Big Push for Renewable Energy in India: What Will Drive It?" Bulletin of the Atomic Scientists, 71(4), July/August, 31–42. Available at: http://bos.sagepub.com/content/71/4/31.abstract.

7. Mohit Sharma and Arunabha Ghosh (2015) "Imagining Smart Cities in India," in What Does India Think? Edited by François Godement, 71–75. London: European Council on Foreign Relations. Available at: http://www.ecfr.eu/page/-/ECFR145_WDIT.pdf

8. Abhishek Jain, Sudatta Ray, Karthik Ganesan, Michael Aklin, Chao-Yo Cheng and Johannes Urpelainen (2015) "Access to Clean Cooking Energy and Electricity: Survey of States," September. New Delhi: Council on Energy, Environment and Water. Available at: http://ceew.in/pdf/CEEW-ACCESS-Report-29Sep15.pdf

9. Karthik Ganesan and Rajeev Vishnu (2014) "Energy Access in India—Today, and Tomorrow," CEEW Working Paper 2014/10, June. Available at: http://ceew.in/pdf/CEEW-Energy-Access-in-India-Today-and-Tomorrow-1Jul14.pdf

10. Abhishek Jain, Poulami Choudhury, and Karthik Ganesan (2015) "Clean, Affordable and Sustainable Cooking Energy for India: Possibilities and Realities beyond LPG," February. New Delhi: Council on Energy, Environment and Water. Available at: http://ceew.in/pdf/ceew-clean-affordable-and-sustainable-cooking.pdf

11. Arunabha Ghosh and Karthik Ganesan (2015) "Rethink India's Energy Strategy" Nature 521, 14 May, 156–157. Available at: http://www.nature.com/news/policy-rethink-india-s-energy-strategy-1.17508 Also, Abhishek Jain, Shalu Agrawal, and Karthik Ganesan (2014) "Rationalising Subsidies, Reaching the Underserved: Improving Effectiveness of Domestic LPG Subsidy and Distribution in India," November. Available at: http://ceew.in/pdf/CEEW-Rationalising-LPG-Subsidies-Reaching-the-Underserved-5Dec14.pdf

12. FICCI (2013). Power Transmission—The Real Bottleneck. New Delhi: FICCI.

13. The subdued outlook is primarily a result of the higher priced gas (based on the current outlook on prices). Imported gas is not able to compete against coal except in an extremely stringent regime to cut carbon or at exceptionally high carbon prices.

14. PPAC (2014) Forecast and Analysis, PPAC. Retrieved 2014, from Petroleum Planning and Analysis Cell: http://ppac.org.in/

15. IIR (2010) India Infrastructure Report—Infrastructure Development in a Low Carbon Economy. New Delhi: Oxford University Press.

16. This portion on infrastructure lock-in has been adapted from the framing India Infrastructure Report (2010).

17. Vaibhav Chaturvedi, Vaibhav Gupta, Nirmalya Choudhury, Sonali Mittra, Arunabha Ghosh, and Rudresh Sugam (2014) State of Environmental Clearances in India: Procedures, Timelines and Delays across Sectors and States, October. New Delhi: Council on Energy, Environment and Water.

18. M.R. Madhavan, (2013) "Land Acquisition Process Can Take 50 Months." PRS Legislative Research, 06 September. Available at: http://www.prsindia.org/theprsblog/?tag=land-acquisition; accessed on 17 July 2015.

19. P. Nambiar (2013, December 02) GAIL's $3-billion pipelines on hold for want of buyers. From The financial Express: http://www.financialexpress.com/news/gail-s-3billion-pipelines-on-hold-for-want-of-buyers/1201985/0; accessed in June 2014.

20. M. Lakhamraju (2010) Manpower Crunch in Energy Sector. From Great Lakes Management Institute. Available at: http://www.greatlakes.edu.in/gurgaon/sites/default/files/Manpower_Crunch_in_Energy_Sector.pdf; accessed on 15 July 2015.

21. TERI 2015 and Bhushan 2010. TERI analysis based on data available from the Central Electricity Authority.

22. http://mnre.gov.in/mission-and-vision-2/achievements/

23. US Department of Energy.

24. http://www.cseindia.org/userfiles/39-56%20Steel(1).pdf; accessed on 13 May 2015.

25. World Bank (2014) "World Development Indicators." Available at: http://data.worldbank.org/indicator/NE.TRD.GNFS.ZS?page=2

26. CEEW analysis based on the assumptions around India's oil demand in 2030 (6 mb/d), import dependency (80%) and growth in global oil trade (at an annual average growth rate of 2%).

27. BP (2015) Statistical Review of World Energy 2015. Available at: http://www.bp.com/en/global/corporate/about-bp/energy-economics/statistical-review-of-world-energy/statistical-review-downloads.html

28. Ibid.

29. David Steven and Arunabha Ghosh (2014) "Materials, Markets, Multilateralism: A Strategic Approach to India's Resource Challenges," in The Politics of Strategic Resources: Energy and Food Security Challenges in the 21st Century, edited by David Steven, Emily O'Brien and Bruce Jones. Washington, D.C.: Brookings Press, p. 42.

30. Authors' analysis based on the total oil consumption and production data for India from US Energy Information Administration (2014) International Energy Statistics and Short-Term Energy Outlook, June.

31. Petroleum Planning and Analysis Cell (2015) Snapshot of India's Oil and Gas Data. Available at: http://ppac.org.in/WriteReadData/Reports/201505210643441963848SnapshotofIndia'sOil&Gasdata-April2015.pdf

32. McKinsey (2014) "India Towards Energy Independence 2030." Available at: http://www.mckinsey.com/~/media/mckinsey%20offices/india/pdfs/india_towards_energy_independence_2030.ashx

33. Central Electricity Authority (2015) "Monthly Executive Summary Report, March 2015." Available at: http://www.cea.nic.in/reports/monthly/executive_rep/Mar15.pdf

34. PNGRB (2013) "Vision 2030—Natural Gas infrastructure in India", Available at: http://www.pngrb.gov.in/Hindi-Website/pdf/vision-NGPV-2030-06092013.pdf

35. Press Information Bureau (2015) "Prime Minister inaugurates Urja Sangam 2015—India's Global Hydrocarbon Summit," 27 March. Available at http://pib.nic.in/newsite/PrintRelease.aspx?relid=117787; accessed 7 October 2015.

36. Press Trust of India (2015) "India Should Buy Overseas Oil and Gas Assets Aggressively: Dharmendra Pradhan," The Economic Times, 9 September. Available at: http://articles.economictimes.indiatimes.com/2015-09-09/news/66363662_1_oil-minister-dharmendra-pradhan-strategic-oil-reserves-oil-import-bill; accessed 7 October 2015.

37. ET Bureau (2015) "India Asks OPEC to Stop Charging Premium from Asian Buyers," Economic Times, 3 June. Available at: http://economictimes.indiatimes.com/industry/energy/oil-gas/india-asks-opec-to-stop-charging-premium-from-asian-buyers/articleshow/47531587.cms; accessed 25 July 2015.

38. Himangshu Watts (2015) "Asian Premium' a Misconception, Ready for Big-Ticket Investment in India: Saudi Oil Minister Al-Naimi,' The Economic Times, 5 October. Available at: http://economictimes.indiatimes.com/industry/energy/oil-gas/asian-premium-a-misconception-ready-for-big-ticket-investment-in-india-saudi-oil-minister-al-naimi/articleshow/49221131.cms; accessed 9 October 2015.

39. Adi Imsirovic and Tilak Doshi (2012) "Exploring the Asian Premium in Crude Oil Markets," Journal of Energy Security, 21 November. Available at: http://ensec.org/index.php?option=com_content&view=article&id=393:exploring-the-asian-premium-in-crude-oil-markets&catid=130:issue-content&Itemid=405; accessed 9 October 2015.

40. http://www.iea.org/topics/energysecurity/

41. http://www.energy.gov/mission

42. www.bmwi.de/English/Redaktion/Pdf/germanys-new-energy-policy

43. http://www.meti.go.jp/english/press/data/20100618_08.html

44. http://www.greengrowth.go.kr/?page_id=42450

45. Jian Zhang (2011), "China's Energy Security: Prospects, Challenges, and Opportunities," The Brookings Institution. Available at: http://www.brookings.edu/research/papers/2011/07/china-energy-zhang

46. Arunabha Ghosh (2006) "Pathways Through Financial Crisis: India" Global Governance 12(4): 413–429.

47. Sergio Peçanha (2013) "Understanding the Deal With Iran," The New York Times. Available at: http://www.nytimes.com/interactive/2013/11/24/world/middleeast/Understanding-the-Deal-With-Iran.html?_r=5&

48. Sanctions allowed the Obama administration to make exemptions to countries that "significantly" reduce their volume of purchases of Iranian crude oil, determined on a case-to-case basis.

49. UN Comtrade Database, http://comtrade.un.org/data/

50. Ibid.

51. http://www.e-ir.info/2012/04/29/indias-approach-to-sanctions-on-iran/

52. http://in.reuters.com/article/2013/03/08/iran-india-imports-idINDEE92707620130308

53. http://www.reuters.com/article/2012/05/21/india-mrpl-iran-idUSL4E8GL3MI20120521

54. http://uk.reuters.com/article/2012/04/17/uk-iran-oil-insurance-idUKLNE83G00G20120417

55. BP Statistical Review of World Energy.

56. India National Energy Model bottom-up projections for demand yield.

57. BP (2015) Energy Outlook 2035, February.

58. Maplecroft (Bath, October 21, 2010) "Big Economies of the Future—Bangladesh, India, Philippines, Vietnam, and Pakistan—Most at Risk from Climate Change." Available at: http://maplecroft.com/about/news/ccvi.html

59. David King, Daniel Schrag, Zhou Dadi, Qi Ye and Arunabha Ghosh (2015) Climate Change: A Risk Assessment. London: UK Foreign and Commonwealth Office, July.

60. UNFCCC (2009) "Copenhagen Accord," FCCC/CP/2009/11/Add.1. Available at: http://unfccc.int/resource/docs/2009/cop15/eng/11a01.pdf; accessed 9 October 2015.

61. ONGC Videsh Ltd., "Annual Report 2013-14" Of OVL assets, 13 are producing; 4 are under development, 14 are in the exploration phase and two are pipeline projects.

62. http://www.business-standard.com/article/companies/state-owned-oil-firms-to-bid-together-for-lng-blocks-abroad-114100200957_1.html

63. http://www.business-standard.com/article/companies/videocon-and-bpcl-find-more-oil-in-brazil-s-sergipe-basin-115010900863_1.html

64. http://www.oil-india.com/CAddress.aspx

65. articles.economictimes.indiatimes.com/2015-04-13/news/61103185_1_golfinho-oil-india-ltd-rovuma-area

66. http://www.livemint.com/Companies/1ZGLXJjKJeMfoS8VEygptK/RIL-plans-to-invest-2-billion-in-US-shale-gas-assets.html

67. http://economictimes.indiatimes.com/industry/energy/oil-gas/ril-and-pioneer-natural-resources-company-close-deal-to-sell-efs-midstream/articleshow/47996538.cms

68. http://www.livemint.com/Companies/TadepRPclULnEYKnpYm35I/RIL-BP-give-up-2-more-oil-and-gas-blocks-tally-down-to-4.html

69. http://economictimes.indiatimes.com/industry/indl-goods/svs/metals-mining/singareni-seeks-aid-coal-videsh-to-buy-foreign-coal-assets/articleshow/40294815.cms

70. This includes investment plans of $6.5 billion, $10 billion and $15.4 billion by Coal India Ltd., GVK group and Adani group, respectively. Reference: http://www.business-standard.com/article/companies/adani-gvk-lanco-seek-to-cut-stake-in-australian-coal-mines-114050801185_1.html

71. http://www.rediff.com/money/report/india-inc-staring-at-huge-losses-from-overseas-arms/20150608.htm

72. Dev Chatterjee (2014) "Adani, GVK, Lanco Seek to Cut Stake in Australian Coal Mines," Business Standard, 9 May. Available at: www.business-standard.com/article/companies/adani-gvk-lanco-seek-to-cut-stake-in-australian-coal-mines-114050801185_1.html; accessed 9 October 2015.

73. http://articles.economictimes.indiatimes.com/2015-04-06/news/60865955_1_foreign-acquisitions-additional-coal-coal-india

74. IEA (2014) "Partner Country Series—Update on Overseas Investments by China's National Oil Companies." Available at: https://www.iea.org/publications/freepublications/publication/partner-country-series---update-on-overseas-investments-by-chinas-national-oil-companies.html

75. IEA (2011) "Overseas Investments by Chinese National Oil Companies." Available at: https://www.iea.org/publications/freepublications/publication/overseas-investments-by-chinese-national-oil-companies.html

76. IEA (2011) "Overseas Investments by Chinese National Oil Companies." Available at: https://www.iea.org/publications/freepublications/publication/overseas-investments-by-chinese-national-oil-companies.html

77. IEA (2014) "Partner Country Series—Update on Overseas Investments by China's National Oil Companies." Available at: https://www.iea.org/publications/freepublications/publication/partner-country-series---update-on-overseas-investments-by-chinas-national-oil-companies.html

78. China National Petroleum Corporation (CNPC), China Petroleum and Chemical Corporation (Sinopec Group), and National Offshore Oil Corporation (CNOOC) are the largest state-owned oil companies. PetroChina is CNPC's publicly-listed arm; Sinopec is the traded part of Sinopec Group. State-owned Sinochem, CITIC Energy and private companies like YanChang Petroleum have also invested in overseas assets.

79. http://www.ongcvidesh.com/news/press-releases/

80. http://www.reuters.com/article/2014/10/27/india-ongc-idUSL4N0SM2FR20141027

81. Authors' analysis

82. Authors' analysis

83. http://www.thehindu.com/business/Economy/india-loses-kashagan-oil-field-to-china/article4873734.ece

84. Wu Fuzuo (2010) Yazhou Nengyuan Xiaofeiguo jian de Nengyuan Jingzheng yu Hezuo: Yizhong Boyi de Fenxi (Energy Cooperation and Competition among Asian Energy Consuming Countries: A Game Theory Analysis) Shanghai: Shanghai People's Publishing House, Chapter 4, "Sino-Indian Autonomous Cooperation in Energy Game."

85. Arunabha Ghosh (Forthcoming) "India's Resource Nexus: What Scope for Cooperation with China?" in China and India: Towards Cooperation between the Giants of Asia, edited by Kishore Mahbubani, Huang Jing and Kanti Bajpai. London: Routledge.

86. "China, India Sign Energy Agreement," China Daily, 13 January 2006. Available at: http://www.chinadaily.com.cn/english/doc/2006-01-13/content_511871.htm

87. Rakesh Sharma (2012) "India, China to Explore Energy Assets," The Wall Street Journal, 19 June. Available at: http://blogs.wsj.com/dealjournalindia/2012/06/19/ongc-cnpc-to-renew-pact-on-cooperation/

88. Lydia Powell (2012) "Do India's Equity Oil Investments Make Sense?" Energy News Monitor, VIII/43, New Delhi: Observer Research Foundation, 10 April. Available at: http://www.observerindia.com/cms/sites/orfonline/modules/enm-analysis/ENM-ANALYSISDetail.html?cmaid=35815&mmacmaid=35813

89. S. Anilesh Mahajan (2012) "World Wide Woe – ONGC Videsh's Overseas Woes: Could the Problems have been Avoided?" Business Today, 19 August. Available at: http://businesstoday.intoday.in/story/overseas-problems-of-ongc-videsh-other-oil-companies/1/186797.html; Rakesh Sharma (2012) "ONGC to Continue Exploration in South China Sea," Wall Street Journal, 19 July. Available at: http://online.wsj.com/article/SB10000872396390444464304577536182763155666.html

90. Press Trust of India (2015) "ONGC Videsh Ltd Seeks Autonomy to Invest $1 Billion to Speed Up Acquisition of Overseas Fields," 17 June. Available at: http://economictimes.indiatimes.com/industry/energy/oil-gas/ongc-videsh-ltd-seeks-autonomy-to-invest-1-billion-to-speed-up-acquisition-of-overseas-fields/articleshow/47705937.cms?intenttarget=no; accessed 25 July 2015.

91. Including crude oil and product carriers.

92. Indian Petroleum and Natural Gas Statistics (2013–2014).

93. http://blogs.platts.com/2014/08/18/china-oil-tankers/

94. http://blogs.platts.com/2014/08/18/china-oil-tankers/

95. Andrew Erickson and Gabe Collins (2007) "Beijing's Energy Security Strategy: The Significance of a Chinese State-Owned Tanker Fleet," Orbis 51(4), 665–684. Available at: http://www.sciencedirect.com/science/article/pii/S0030438707000816; accessed 9 October 2015.

96. Andrew Stevens (2015) "China is Hoarding Cheap Oil in a Fleet of Supertankers," CNN Money, 4 June. Available at: http://money.cnn.com/2015/06/04/news/economy/china-oil-supertanker-opec/; accessed 9 October 2015.

97. Analysis using data from UN Comtrade Database, http://comtrade.un.org/data/

98. See note 95.

99. http://www.dnaindia.com/money/report-shipping-corp-lines-up-big-liquefied-natural-gas-push-1994986

100. Wendy Laursen (2015) "India and Korea Boost Shipbuilding Ties," The Maritime Executive, 19 May. Available at: http://maritime-executive.com/article/india-and-korea-boost-shipbuilding-ties

101. IGU (2014), "World LNG report 2014," International Gas Union.

102. PNGRB (2013) "Vision 2030—Natural Gas infrastructure in India." Available at: http://www.pngrb.gov.in/Hindi-Website/pdf/vision-NGPV-2030-06092013.pdf; accessed 9 October 2015.

103. ICRA Research Services (2015) Industry Review—Indian Gas Utilities, July. Available at: www.icra.in/Files/ticker/SH-2015-Q3-1-ICRA-Gas%20Utilities.pdf

104. See note 101.

105. http://www.dnaindia.com/money/report-hiranandani-gets-green-nod-for-lng-terminal-2052034

106. Sanjeev Chaudhari (2015), "New Delhi to Seek Revival of Iran-Pak-India Gas Pipeline," The Economic Times, 28 July. Available at: http://economictimes.indiatimes.com/articleshow/48244960.cms?utm_source=contentofinterest&utm_medium=text&utm_campaign=cppst

107. Amitav Rajan (2015) "Stuck in The Pipeline: A $4-Billion Deep-Sea Gas Project," The Indian Express, 13 May. Available at: http://indianexpress.com/article/india/india-others/stuck-in-the-pipeline-a-4-billion-deep-sea-gas-project/

108. Oscar Nkala (2015) "India Developing Network of Coastal Radars," Defense News, 20 March. Available at: http://www.defensenews.com/story/defense/naval/2015/03/20/india-seychelles-coastal-radar-china-modi-indian-ocean/25084237/

109. http://www.ndtv.com/india-news/in-south-china-sea-row-top-us.-commander-roots-for-india-743991

110. National University of Singapore (2014) ISAS NUS Working Paper No. 185, 26 March, p. 14.

111. http://www.indiastrategic.in/navy.htm

112. http://www.dw.de/naval-buildup-reflects-indias-ambition-to-project-power/a-18275292

113. CASS-India "Rising Tension: India Asserts its Presence in South China Sea," Centre for Asian Strategic Studies—India. Available at: http://www.cassindia.com/inner_page.php?id=27&&task=diplomacy

114. http://www.dw.de/naval-buildup-reflects-indias-ambition-to-project-power/a-18275292

115. http://www.businessinsider.in/Indian-Navys-Blue-Water-journey-the-road-ahead/articleshow/47515574.cms

116. See note 110.

117. http://defencyclopedia.com/2015/07/03/dragon-vs-elephant-part-2-indian-navys-role-in-the-indian-ocean/

118. http://defencyclopedia.com/2015/07/03/dragon-vs-elephant-part-2-indian-navys-role-in-the-indian-ocean/

119. Premvir Das (2015) "Navy to Navy," Business Standard, 16 June. Available at: http://www.business-standard.com/article/opinion/premvir-das-navy-to-navy-115061501172_1.html#.VYjh7W92ZSk.twitter; accessed 24 July 2015.

120. India has also developed plans to build a strategic uranium reserve, which could have enough fuel to last 5–10 years. Andrew Topf (2015) "India to Create Strategic Uranium Reserve," 19 July. Available at: http://www.mining.com/india-to-create-strategic-uranium-reserve/, accessed 25 July 2015.

121. http://www.isprlindia.com/aboutus-2.asp

122. ISPRL (2014) "Annual Report 2013–14." ISPRL.

123. http://in.reuters.com/article/2015/03/30/india-energy-spr-idINKBN0MQ1N220150330

124. Petroleum Planning and Analysis Cell (2015) Snapshot of India's Oil and Gas Data. Available at: http://ppac.org.in/WriteReadData/Reports/201505210643441963848SnapshotofIndia'sOil&Gasdata-April2015.pdf

125. http://www.eia.gov/dnav/pet/pet_move_neti_dc_NUS-Z00_mbblpd_a.htm

126. http://www.bloomberg.com/news/articles/2013-05-14/china-seen-boosting-emergency-oil-storage-capacity-iea-says

127. http://www.iea.org/netimports/

128. http://www.ibtimes.co.in/india-build-5-mt-underground-oil-storage-reserves-by-october-622063

129. This commitment can be met through both stocks held exclusively for emergency purposes and stocks held for commercial or operational use, including stocks held at refineries, at port facilities and in tankers in ports.

130. Revised cost estimates of the 5.33 MMT SPR capacity nearing completion. Source: ISPRL (2014) "Annual Report 2013–14".

131. Anshuman Mainkar (2015) "Building on Strategic Reserve," The Hindu, 28 May. Available at: http://www.thehindu.com/opinion/op-ed/article7252347.ece, accessed 24 July 2015.

132. http://in.reuters.com/article/2014/11/17/india-refiners-idINKCN0J10FN20141117

133. http://www.ibtimes.co.uk/india-considers-foreign-investment-build-12-million-ton-oil-reserves-1492964

134. http://www.livemint.com/Home-Page/7csswmMpXS6Z3KaeCNmroJ/Oman-may-invest-in-Indian-crude-reserves.html

135. http://energy.gov/fe/services/petroleum-reserves/strategic-petroleum-reserve/spr-quick-facts-and-faqs

136. http://www.bloomberg.com/news/articles/2013-05-14/china-seen-boosting-emergency-oil-storage-capacity-iea-says

137. Government of India (2006) "Crisis Management: From Despair to Hope," Second Administrative Reforms Commission, Third Report, September. Available at: http://arc.gov.in/3rdreport.pdf; accessed 9 October 2015.

138. https://www.iea.org/topics/energysecurity/subtopics/respondingtomajorsupplydisruptions/

139. https://www.iea.org/topics/energysecurity/subtopics/stockholdingstructure/

140. https://www.iea.org/topics/energysecurity/subtopics/stockholdingstructure/

141. Arunabha Ghosh (2015) "Shifts and risks in energy," Business Standard, 19 May. Available at: http://www.business-standard.com/article/opinion/arunabha-ghosh-shifts-and-risks-in-energy-115051801418_1.html, accessed 25 July 2015.

142. David Steven and Arunabha Ghosh (2014) "Materials, Markets, Multilateralism: A Strategic Approach to India's Resource Challenges", in The Politics of Strategic Resources: Energy and Food Security Challenges in the 21st Century, edited by David Steven, Emily O'Brien and Bruce Jones. Washington, D.C.: Brookings Press, p. 57.

143. Ann Florini (2011) "The International Energy Agency in Global Energy Governance," Global Policy 2 (1, September): 40–50.

144. Arunabha Ghosh (2011) "Seeking Coherence in Complexity? The Governance of Energy by Trade and Investment Institutions," Global Policy, 2 (Special Issue), September, 106–119.

145. David G. Victor and Linda Yueh (2010) "The New Energy Order: Managing Insecurities in the Twenty-first Century," Foreign Affairs, 89/1, January/February, 71–72.

146. Arunabha Ghosh, with Himani Gangania (2012) Governing Clean Energy Subsidies: What, Why, and How Legal? Geneva: International Centre for Trade and Sustainable Development.

147. Arunabha Ghosh (2015) "The Big Push For Renewable Energy in India: What Will Drive It?" Bulletin of the Atomic Scientists, 71(4), July/August, 31–42.

148. Ibid.

149. Government of India (2015) "India's Intended Nationally Determined Contribution: Working Towards Climate Justice," 2 October. Available at: http://www4.unfccc.int/submissions/INDC/Published%20Documents/India/1/INDIA%20INDC%20TO%20UNFCCC.pdf; accessed 9 October 2015.

150. Mohit Sharma and Arunabha Ghosh (2015) "Imagining smart cities in India," in *What does India think?* Edited by François Godement, 71–75. London: European Council on Foreign Relations. Available at: http://www.ecfr.eu/page/-/ECFR145_WDIT.pdf

151. Including Saudi Arabia, Russia, the United States, China and Norway.

152. Press Information Bureau 2015. Marginal fields are where hydrocarbon discoveries were made by ONGC and OIL, but not monetized due to, for example, isolated locations, small reserves, high development costs and technological constraints.

153. The KG-D6 basin is the Dhirubhai-6, where Reliance Industries discovered the biggest gas site in India. The field is spread across the Krishna and Godavari Basins near the coast of Andhra Pradesh. The basin could have estimated reserves of 10 trillion cubic feet (TCF) of gas (Jayaswal & Banerjee, 2009).

154. The NELP was formulated in 1997 (coming into effect in 1999) with the aim of attracting risk capital (domestic and international), technologies and geological concepts and practices to explore oil and gas resources (Ministry of Petroleum and Natural Gas, 2015).

155. http://pib.nic.in/newsite/PrintRelease.aspx?relid=117888

156. PIB 2015. Two CIL subsidiaries—Mahanadi Coalfields Ltd. and the South Eastern Coalfields Ltd.—are expected to play pivotal roles, with contributions of 250 MT and 240 MT, respectively.

157. As it is for oil and gas. For example, E&P may require around 70 clearances (PETROFED, cited in Soni and Chatterjee, 2014). Some offshore blocks have even been declared "no-go" areas by the Ministry of Defence after blocks have been allocated and the petroleum exploration licence granted.

158. Manohar Lal Sharma vs The Principal Secretary & Ors, 2014.

159. TERI analysis, REF scenario.

160. A description is in the annex to Chapter 1.

161. Cooking options included improved cook-stoves and LPG gas-stoves; lighting options included tubelights, CFL bulbs, LED bulbs and solar lanterns.

162. Thus, the cooking energy basket remains dominated by traditional cooking (biomass) options.

163. The analysis assumes that the pithead price of underground coal is Rs. 3,724 a tonne in 2014, while the cost, insurance, freight price of imported coal is Rs. 3,702 a tonne.

164. The dissent note from the Secretary, MoPNG, is annexed to the Report of the Committee.

165. Also stated in the Chawla Committee Report of 2011.

166. http://www.pngrb.gov.in/the-act.html

167. The Bill awaits presidential consent.

168. The Hindu Business Line 2014.

169. This separation is a condition of the India–US nuclear accord of 2008 and the safeguard agreement of the International Atomic Energy Agency (Application of Safeguards to Civilian Nuclear Facilities) of 2009.

170. World Energy Outlook 2012, IEA.

About the Authors

Suman Bery is Chief Economist, Shell International, based in The Hague, Netherlands. He assumed this position on 1 February 2012.

Mr Bery served as Director-General (Chief Executive) of the National Council of Applied Economic Research (NCAER), New Delhi, from January 2001 to March 2011. NCAER is one of India's leading independent policy research institutions. He then served as Country Director-India Central, the International Growth Centre (IGC). The IGC is a research initiative of UK Aid in partnership with the London School of Economics and the University of Oxford. In this capacity he was responsible for setting up the IGC's New Delhi office in partnership with the Indian Statistical Institute's Delhi centre.

Prior to NCAER, Mr Bery was with the World Bank in Washington, D.C. From 1992 to 1994, on leave from the World Bank, Mr Bery worked as Special Consultant to the Reserve Bank of India, Bombay, where he advised the Governor and Deputy Governors on financial sector policy, institutional reform, and market development and regulation.

Mr Bery completed his undergraduate work at Magdalen College, University of Oxford in Philosophy, Politics and Economics and holds a Master of Public Affairs (MPA) degree from the Woodrow Wilson School of Public and International Affairs, Princeton University.

Arunabha Ghosh, PhD, has been CEO of the Council on Energy, Environment and Water (CEEW) since its founding in August 2010. CEEW has been consistently ranked (third year running) as South Asia's leading policy research institution across several categories. With work experience in 37 countries and having previously worked at Princeton, Oxford, UNDP (New York) and WTO (Geneva), co-author of four books and dozens of research papers and reports, Arunabha advises governments, industry, civil society and international organizations around the world. He is a World Economic Forum *Young Global Leader*, Asia Society's *Asia 21 Young Leader*, and fellow of the *Aspen Global Leadership Network*.

Dr Ghosh was invited by the Government of France as a *Personnalité d'Avenir* to advise on the COP21 climate negotiations. He has been actively involved, since inception, in developing the strategy for and supporting activities related to the International Solar Alliance, which was launched by the governments of India and France in November 2015. He serves on the Executive Committee of the India-U.S. PACEsetter Fund, which invests in promising decentralized energy solutions. He has presented to heads of state and legislatures across the world; and is a member of Track II dialogues with seven countries.

Arunabha's co-authored essay "Rethink India's energy strategy" in *Nature*, the world's most cited scientific journal, was selected as one of 2015's 10 most influential essays. Widely published, he is most recently a co-author of *Climate Change: A Risk Assessment* (2015) and *Human Development and Global Institutions* (Routledge, 2016). Another forthcoming book is *The Palgrave Handbook of the International Political Economy of Energy* (2016). He has been an author of three UNDP *Human Development Reports*. He writes a monthly column in the *Business Standard*, and has hosted a documentary on water in Africa and one on energy in India.

Dr Ghosh is a founding board member of the Clean Energy Access Network (CLEAN) and is a board member of the International Centre for Trade & Sustainable Development, Geneva. He holds a doctorate from the University of Oxford (Clarendon Scholar; Marvin Bower Scholar), an M.A. (First Class) in Philosophy, Politics and Economics (Balliol College, Oxford; Radhakrishnan Scholar); and topped Economics from St. Stephen's College, Delhi.

Ritu Mathur, PhD, has been leading the Energy Policy Modelling and Scenario Building activities at TERI over the last two decades and is currently also associated with TERI University as Professor in the Department of Energy and Environment.

An economist by training and with a PhD in Energy Science from Kyoto University, Japan, her work largely focuses on addressing policy and regulatory aspects related to the energy sector, examining the potentials and challenges to cleaner energy choices while addressing energy security and development-related considerations of developing countries, cost–benefit analysis and evaluation of synergies and trade-offs of alternative technological and policy pathways, etc.

She has led several national and international interdisciplinary projects and has authored several papers, reports and books related with energy use and its implications on the environment at the local and global levels in order to inform and influence various stakeholders in decision-making.

Dr Mathur has been associated as a member of several committees including the Expert Group on Low Carbon Strategies for Inclusive Growth, Steering Committee on Energy Sector for India's Five Year Plans, Sustainable Growth Working Group of the U.S.-India Energy Dialogue led by NITI Aayog (formerly Planning Commission of India), etc. She was also a Lead Author in Working Group III of the IPCC Fifth Assessment Report, and a Reviewer in IPCC AR4.

Subrata Basu is Consultant in Finance and Strategy team of Shell based in Bengaluru. He has been working in various energy sector companies in India and with Shell for the last 5 years. His areas of interest include energy sector policies, new technology development in the energy sector, energy market strategies for organizations, etc. He has previously worked with organizations such as Larsen & Tubro and PwC in various operational and strategic roles. He has contributed to international and national journals and conferences. In Shell, his current role includes international policy review, internal business and investment planning. He enjoys travelling, music and sports.

Karthik Ganesan is a Senior Research Associate at the Council on Energy, Environment and Water (CEEW), India. As a member of the team at CEEW his research focus includes the development of long-term energy scenarios for India (based on an in-house cost-optimization model) and energy efficiency improvements in the industrial sector in India. Linked to his work in industrial efficiency is his role as the principal investigator in an effort to identify critical mineral resources required for India's manufacturing sector. He also leads a civil society effort to assess greenhouse gas emissions from the industrial sector to understand the contribution of the sector to national emissions inventory. In addition, he supports ongoing work in the areas of energy access indicators for rural Indian households and carried out a first-of-a-kind evaluation of the impact of industrial policies on the renewable energy sector in India. Prior to his association with CEEW, he has worked on an array of projects in collaboration with various international institutions, with a focus on low-carbon development and energy security.

His published (and under review) works include Rethink India's Energy Strategy (*Nature*, Comment); The Co-location Opportunities for Renewable Energy and Agriculture in North-western India: Trade-offs and Synergies (Applied Energy, American Geophysical Union); Valuation of Health Impact of Air Pollution from Thermal Power Plants (ADB); Technical Feasibility of Metropolitan Siting of Nuclear Power Plants (NUS) and Prospects for Carbon Capture and Storage in SE Asia (ADB). His role as a research assistant at a graduate level focused on the linkages between electricity consumption and sectoral economic growth using a time-series approach.

Karthik has a Master of Public Policy from the Lee Kuan Yew School of Public Policy at the National University of Singapore (NUS). His prior educational training resulted in an MTech in Infrastructure Engineering and a BTech in Civil Engineering from the Indian Institute of Technology Madras.

Rhodri Owen-Jones is an Energy Analyst in Corporate Strategy and Planning, having joined Shell in 2008. In his current position, Rhodri is heavily involved in the quantification and modelling of the Shell New Lens Scenarios, as well as communicating the resulting work to a wider global audience. His work also includes modelling global long-term energy supply and demand, analysing and advising senior leadership on short-term oil and gas market developments as well as managing a joint research project on future Indian energy pathways. He is also Business Advisor to the Executive Vice President of Strategy in the RDS Group. Rhodri has previously worked in Production Engineering at NAM BV as well as coordinating and running a Europe-wide benchmarking exercise of Shell's European assets. He holds a Masters in Mechanical Engineering having graduated from the University of Bath in 2008 with first-class honours.